47656

MW

MYCOTOXINS IN DAIRY PRODUCTS

MYCOTOXINS
IN DAIRY PRODUCTS

Edited by

HANS P. VAN EGMOND

*National Institute of Public Health & Environmental Protection,
Bilthoven, The Netherlands*

ELSEVIER APPLIED SCIENCE
LONDON and NEW YORK

ELSEVIER SCIENCE PUBLISHERS LTD
Crown House, Linton Road, Barking, Essex IG11 8JU, England

Sole Distributor in the USA and Canada
ELSEVIER SCIENCE PUBLISHING CO., INC.
655 Avenue of the Americas, New York, NY 10010, USA

WITH 27 TABLES AND 45 ILLUSTRATIONS

© 1989 ELSEVIER SCIENCE PUBLISHERS LTD

British Library Cataloguing in Publication Data

Mycotoxins in dairy products.
1. Mycotoxins
I. Van Egmond, H. P.
574.2'326

ISBN 1-85166-369-X

Library of Congress Cataloging-in-Publication Data

Mycotoxins in dairy products/edited by H. P. van Egmond.
   p. cm.
   Includes bibliographies and index.
   Contents: Introduction/H. P. van Egmond—Aflatoxin M_1:
occurrence, toxicity, regulation/H. P. van Egmond—
Chromatographic methods of analysis for aflatoxin M_1/
R. D. Stubblefield and H. P. van Egmond—Immunochemical
methods of analysis for aflatoxin M_1/J. M. Fremy and
F. W. Chu—Aflatoxin M_1: stability and degradation/A. E. Yousef
and E. H. Martin—Toxic metabolites from fungal cheese starter
cultures . . ./G. Engel and M. Teuber—Mycotoxigenic fungal
contaminants of cheese and other dairy products/P. M. Scott.
   ISBN 1-85166-369-X
   1. Mycotoxins. 2. Dairy products—Contamination.
  I. Egmond, H.
P. van (Hans P.)
SF254.M93M93 1989
637—dc20   89-32269
                                  CIP

Phototypesetting by Tech-Set, Gateshead, Tyne & Wear.
Printed in Great Britain at the University Press, Cambridge.

Preface

Since the discovery of the aflatoxins in the early 1960s, thousands of publications about mycotoxins have appeared in many periodicals and books and many conferences relative to the subject have been held. A substantial amount of the mycotoxin research carried out by different scientific disciplines is directly or indirectly related with dairy products. The fact that a book on mycotoxins in dairy products has now appeared is evidence of the significant stream of information which has become available. It also emphasizes the many research efforts of research institutes and industries to further improve the high quality of dairy products.

The various chapters in this book are written by experts in the particular fields. I wish to express my sincere appreciation to them for their contributions. I hope the book will provide information to those who want to get an insight into the various aspects of mycotoxins in dairy products.

HANS P. VAN EGMOND

Contents

Preface v

List of Contributors ix

1. Introduction 1
   H. P. VAN EGMOND

2. Aflatoxin M_1: Occurrence, Toxicity, Regulation 11
   H. P. VAN EGMOND

3. Chromatographic Methods of Analysis for Aflatoxin M_1 . 57
   R. D. STUBBLEFIELD and H. P. VAN EGMOND

4. Immunochemical Methods of Analysis for Aflatoxin M_1 . 97
   J. M. FRÉMY and F. S. CHU

5. Stability and Degradation of Aflatoxin M_1 127
   A. E. YOUSEF and E. H. MARTH

6. Toxic Metabolites from Fungal Cheese Starter Cultures . 163
   (*Penicillium camemberti* and *Penicillium roqueforti*)
   G. ENGEL and M. TEUBER

vii

7. Mycotoxigenic Fungal Contaminants of Cheese and Other
    Dairy Products 193
        P. M. SCOTT

Index   261

List of Contributors

F. S. CHU
Department of Food Microbiology and Toxicology, Food Research Institute, University of Wisconsin–Madison, 1925 Willow Drive, Madison, Wisconsin 53706, USA

G. ENGEL
Institut für Mikrobiologie, Bundesanstalt für Milchforschung, D-2300 Kiel, Federal Republic of Germany

J. M. FRÉMY
Ministere de l'Agriculture, Direction Generale de l'Alimentation, Services Veterinaires, Laboratoire Central d'Hygiene Alimentaire, 43, rue de Dantzig, F75015 Paris, France

E. H. MARTH
Department of Food Science and The Food Research Institute, University of Wisconsin-Madison, Madison, Wisconsin 53706, USA

P. M. SCOTT
Health and Welfare Canada, Health Protection Branch, Bureau of Chemical Safety, Ottawa, Ontario, Canada K1A OL2

R. D. STUBBLEFIELD
Northern Regional Research Center, 1815 North University Street, Peoria, Illinois 61604, USA

M. TEUBER
Institut für Mikrobiologie, Bundesanstalt für Milchforschung, D-2300 Kiel, Federal Republic of Germany

H. P. VAN EGMOND
National Institute of Public Health and Environmental Protection, Laboratory for Residue Analysis, Bilthoven, The Netherlands

A. E. YOUSEF
Department of Food Science and The Food Research Institute, University of Wisconsin-Madison, Madison, Wisconsin 53706, USA

Chapter 1

Introduction

H. P. van Egmond

*National Institute of Public Health and Environmental Protection,
Bilthoven, The Netherlands*

1. Significance of Mycotoxins 1
2. Contamination of Dairy Products by Mycotoxins 5
    2.1 Carry over
    2.2 Direct contamination

1 SIGNIFICANCE OF MYCOTOXINS

Mycotoxins are secondary metabolites of fungi which are capable of producing acute toxic, carcinogenic, mutagenic, teratogenic and oestrogenic effects on animals at the levels of exposure. Biological conversion products of mycotoxins are also referred to as mycotoxins. The term 'mycotoxin' is derived from the Greek words 'ΜΥΚΗΣ' (fungus) and 'ΤΟΞΙΚΟΝ' (arrow-poison). Toxic syndromes resulting from the intake of mycotoxins by man and animals are known as 'mycotoxicoses'.

Mycotoxicoses have been known for a long time. The first recognized mycotoxicosis was probably ergotism (Tulasne, 1853), a disease, characterized by necrosis and gangrene of the limbs and better known in the Middle Ages in Europe as 'Holy fire'. This disease was caused by the ingestion of grain contaminated with sclerotia of *Claviceps purpurea*, that contained toxic metabolites. Another mycotoxicosis, recognized to have seriously injured human populations is Alimentary Toxic Aleukia (ATA) (Mayer, 1953). The symptoms of ATA in man take on many forms, including leukopenia, necrotic lesions of the oral cavity, the oesophagus and stomach, sepsis, haemorrhagic diathesis and depletion of the bone marrow. The disease was induced by ingesting overwintered mouldy grain and occurred in many areas in Russia, especially during

1

World War II. It has been reported (Joffe, 1965) that in 1944 more than 10% of the population in certain districts of Russia was affected and many fatalities occurred. The fungi responsible for these accidents belong to the genera *Fusarium* and *Cladosporium*. In Japan, toxicity associated with yellow coloured mouldy rice was a problem, especially after World War II, when rice had to be imported from various countries (Kinosita & Shikata, 1965). The ingestion of 'yellow rice' by man caused vomiting, convulsions and ascending paralysis. Death could also occur within 1–3 days after the appearance of the first signs of the disease. The toxin-producing fungi in yellow rice belong to the genus *Penicillium*.

Despite the forementioned examples of mycotoxin-caused diseases in man, mycotoxicoses remained the 'neglected diseases' (Forgacs & Carll, 1962) until the early 1960s, when this attitude changed drastically due to the outbreak of Turkey X Disease in Great Britain (Stevens *et al.*, 1960). Within a few months more than 100,000 turkeys died, mainly in East Anglia and southern England. In addition, the death of thousands of ducklings and young pheasants was reported (Asplin & Carnaghan, 1961). The problem of Turkey X Disease led to a multidisciplinary approach to investigate the cause of the disease. These efforts were fruitful and the cause of the disease was traced to a toxic factor occurring in the Brazilian groundnut meal which was used as a protein source in the feed of those affected poultry. The toxic factor seemed to be produced by two fungi, *Aspergillus flavus* and *Aspergillus parasiticus*, and hence the name 'aflatoxin' was coined for it, an acronym derived from the name of the first fungus. Further elucidation of the toxic factor demonstrated that the material could be separated chromatographically into four distinct spots (Nesbitt *et al.*, 1962; Hartley *et al.*, 1963). All four components have been given the name 'aflatoxins' in order to identify their generic origin. Distinction of the four substances (aflatoxins B_1, B_2, G_1 and G_2) was made on the basis of their fluorescent colour with subscripts relating to the relative chromatographic mobility. Subsequently, it became clear that the group of aflatoxins consists of many more closely related compounds and that some of these compounds are among the most potent carcinogens known (Wogan, 1973).

In the period following the outbreak of Turkey X disease, a wealth of information about aflatoxins has been produced and many other mycotoxins have also been isolated and characterized. The growing knowledge about the real and potential hazard of mycotoxins to human and animal health led various countries to establish legal measures to control mycotoxin contamination of foodstuffs and animal feedstuffs. At the time of writing, mycotoxin legislation existed in at least 56 countries (Van Egmond, 1987). All these countries had regulations for aflatoxins in food and/or feed. In addition, some had specific regulations for one or more other mycotoxins as well. Of these, patulin, ochratoxin A, deoxynivalenol, T-2 toxin and zearalenone are the more important ones. Regulations also exist for ergot which may contain several toxic alkaloids, such as ergotamine and ergocristine. The chemical structures of these mycotoxins are shown in Fig. 1.

Aflatoxin B_1 is the most notorious mycotoxin, feared for its high toxicity and carcinogenicity. Aflatoxin B_1 occurs particularly in groundnuts, maize and other grains, such as rice, wheat, sorghum and millet. The molecule has a characteristic difuran moiety and a coumarin structure. Both parts of the molecule are suspected to play a role in the carcinogenicity of aflatoxin B_1. When cows are fed with feedstuffs containing aflatoxin B_1, the toxic metabolite aflatoxin M_1 occurs in their milk and subsequently in other dairy products. Aflatoxin M_1 will be treated in detail in various chapters of this book.

Patulin may occur in fruits and fruit juices. The small molecule has a lactone ring as aflatoxin B_1. Patulin has been used in the past as an antibiotic, but later it became known as mycotoxin because it caused haemorrhages and oedema in experimental animals. Nowadays patulin is considered rather an indicator of bad manufacturing practices (use of mouldy raw materials) than a serious threat to human and animal health, as suggested by the results of recent sub-acute and semi-chronic toxicity studies (Speijers & Franken, 1988; Speijers *et al.*, 1988).

Ochratoxin A, a mycotoxin mainly occurring in grains, has been shown to be a potent nephrotoxin in all species of animals tested, including birds, fish and mammals (Krogh, 1977). There is a hypothesis that ochratoxin A is associated with Balkan endemic nephropathy, a renal disease in humans observed in some areas of the Balkan countries (Krogh *et al.*, 1974). Reports indicating the carcinogenicity of ochratoxin A in mice and rats have also been published (Bendele *et al.*, 1985; Boorman 1988).

Deoxynivalenol and T-2 toxin belong to the group of the trichothecenes. The trichothecenes mainly occur in grains. They all have a characteristic epoxy-group in the molecule. These compounds exhibit a wide range of toxic effects in experimental animals including feed refusal, vomiting, diarrhoea and severe intestinal haemorrhage. They have also teratogenic and immunotoxic properties. The occurrence of deoxynivalenol (also referred to as vomitoxin) in feedstuffs for swine may lead to economic damage, because these animals may refuse the

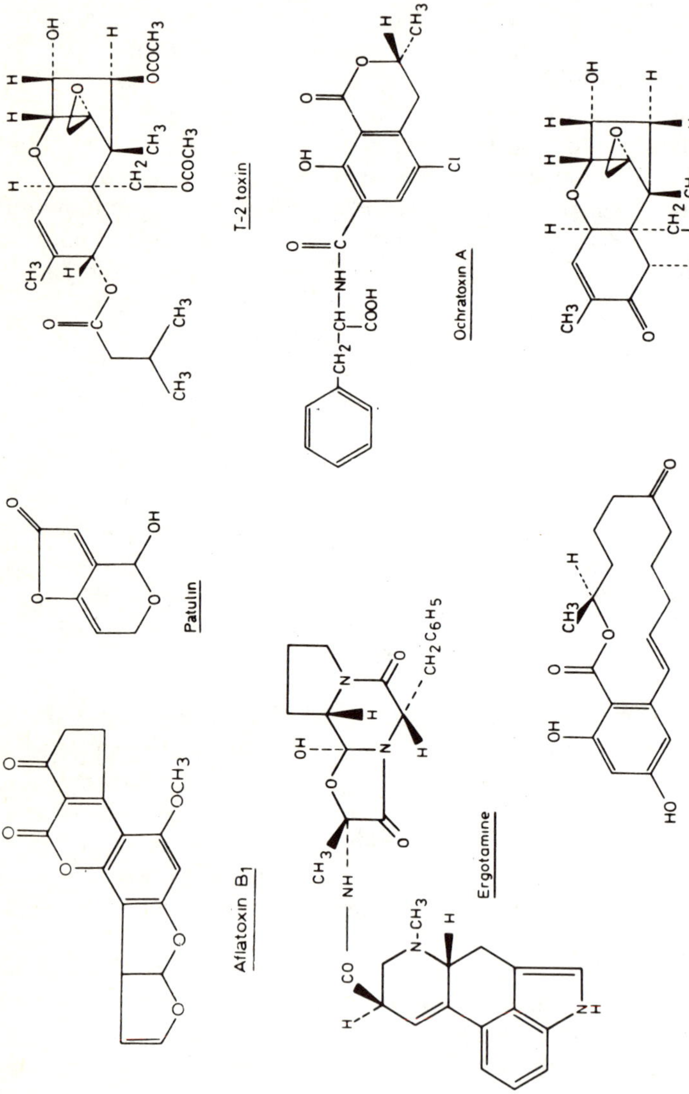

Fig. 1. Chemical structures of some mycotoxins.

feed or show weight loss, suffering from vomiting and diarrhoea. T-2 toxin is a mycotoxin that has attracted particular attention not only as a mycotoxin occurring in foodstuffs but also because of its alleged use in biological warfare in Southeast Asia ('yellow rain') (Anonymous, 1982; Mirocha *et al.*, 1982, 1983).

Zearalenone is a mycotoxin particularly occurring in maize and wheat, and often found together with deoxynivalenol. Zearalenone is related to the anabolic zeranol. It has oestrogenic properties and causes problems with the reproductive organs of farm animals, especially swine.

In addition to the above mentioned examples of mycotoxins for which regulations exist, there are many more mycotoxins known, showing a large variety of chemical structures. Some of them will be discussed in relation to their occurrence in dairy products in other chapters of this book.

2 CONTAMINATION OF DAIRY PRODUCTS BY MYCOTOXINS

Mycotoxin contamination of foods and animal feeds depends highly on environmental conditions that lead to mould growth and toxin production. Data about incidence and levels of contamination are limited by many factors, including the resources to conduct surveys, the availability of laboratory facilities to carry out analyses, the sampling procedure(s) used, the reliability and detectability of the analytical methods used and the capabilities of the analyst(s). Nevertheless numerous publications have appeared about the occurrence of various mycotoxins in foods and animal feeds, since the discovery of the aflatoxins in the early 1960s. Probably no edible substance can be regarded as absolutely free from possible mycotoxin contamination, considering that mycotoxin production can occur in the field, during harvest, processing, storage and shipment of a given commodity.

Among the foodstuffs susceptible to mycotoxin contamination, is the important group of the dairy products. The presence of mycotoxins in dairy products may be the result of:

(1) The contamination of the feedstuffs consumed by dairy cattle (indirect contamination).

(2) The contamination of dairy products by fungi, resulting in the formation of mycotoxins (direct contamination).

2.1 Carry-over

In the early 1960s Allcroft & Carnaghan (1962, 1963) discovered that the intake of aflatoxin-contaminated feed by dairy cows leads to the excretion of a toxic factor in the milk of these cows a few hours after ingestion. The toxic factor that induced the same toxicity syndromes as aflatoxin B_1 when newly hatched ducklings were exposed to it, was named aflatoxin M. Allcroft *et al.* (1966) and Holzapfel *et al.* (1966) elucidated the structure of aflatoxin M which seemed to consist of two structurally related compounds, designated aflatoxins M_1 and M_2, and identified as the 4-hydroxy derivatives of aflatoxins B_1 and B_2 respectively (Fig. 2). Aflatoxin M_1 (AFM_1), the milk metabolite of aflatoxin B_1 (AFB_1), attracted the most attention.

The fact that milk, one of the most important and valuable foods, could harbour aflatoxin(s) initiated a boom of scientific research which has to this date not yet ended. Various disciplines became involved in AFM_1 research. Attention was focussed on toxicity studies of various kinds, on the development of analytical methodology to detect AFM_1 in milk and milk products down to sub $\mu g/kg$ levels, and the production of AFM_1 standards for analytical purposes. Many surveys on the occurrence of the toxin in dairy products were undertaken, as well as studies on its stability, the effects of processing and possibilities for elimination. Various countries established legal limits for AFM_1 in milk as well as for AFB_1 in the feedstuffs for dairy cattle. Check Sample Survey Programmes were organized to help the analyst in testing his ability to determine AFM_1 in powdered milk. Powdered milk reference

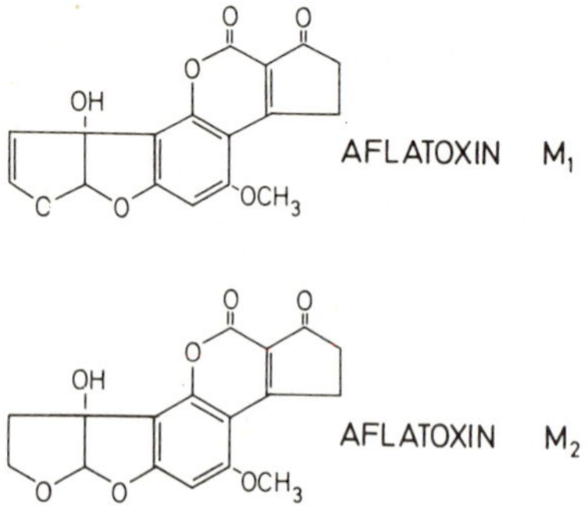

Fig. 2. Chemical structures of aflatoxins M_1 and M_2.

materials certified for their AFM_1 content were developed and made available recently. All these aspects of AFM_1 are discussed in detail in Chapters 2–5.

In addition to aflatoxins there have been studies into the carry-over of other mycotoxins, for instance ochratoxin A, zearalenone (Shreeve *et al.*, 1979), T-2 toxin (Robinson *et al.*, 1979), sterigmatocystin (Kraus, 1978) and deoxynivalenol (Prelusky *et al.*, 1984). However they are relatively rare, and the outcome of these studies does not present a reasonable cause for concern. Therefore these studies will not be reviewed in this book.

2.2 Direct Contamination

The direct contamination of dairy products with mycotoxins may be the result of fungal growth used for fermentation, or unintentional fungal growth. Among the dairy products, cheese is especially susceptible to fungal growth. Moulds that are intentionally grown on cheese are the cheese starter cultures, such as the *Penicillium* species on the French Roquefort and Camembert cheeses. Under certain conditions these fungi are able to produce mycotoxins. The scientific knowledge about the possible formation and occurrence of mycotoxins in intentionally moulded cheeses is reviewed in depth in Chapter 6.

Another possible contamination of dairy products is the accidental occurrence of mould on the products, although good manufacturing practices will often prevent dairy products from getting mouldy. Fortunately, the aflatoxin producers *A. flavus* and *A. parasiticus* do not belong to the most frequent species isolated from dairy products. However many other fungal species may occur, among which is *A. versicolor*, known to be capable of producing versicolorin A and sterigmatocystin, both compounds having carcinogenic properties. The aspects of accidental occurrence of moulds on dairy products in relation to possible mycotoxin formation are comprehensively discussed in Chapter 7.

REFERENCES

Allcroft, R. & Carnaghan, R. B. A. (1962). Groundnut toxicity *Aspergillus flavus* toxin (aflatoxin) in animal products. *Vet. Rec.,* **74**, 863–4.
Allcroft, R. & Carnaghan, R. B. A. (1963). Groundnut toxicity: an examination

for toxin in human products from animals fed toxic groundnut meal. *Vet. Rec.,* **75**, 259–63.

Allcroft, R., Rogers, H., Lewis, G., Nabney, J. & Best, P. E. (1966). Metabolism of aflatoxin in sheep: excretion of the 'milk toxin'. *Nature (London),* **209**, 154–5.

Anonymous (1982). Mycotoxins in South-East Asia. *Nature (London),* **296**, 379–80.

Asplin, F. D. & Carnaghan, R. B. A. (1961). The toxicity of certain groundnut meals for poultry with special reference to their effect on ducklings and chickens. *Vet. Rec.,* **73**, 1215–19.

Bendele, A. M., Carlton, W. W., Krogh, P. & Lillehoj, E. B. (1985). Ochratoxin A carcinogenesis in the (C 57 BL/6J × C3H) F_1 mouse. *J. Natl Cancer Inst.,* **75**, 733–42.

Boorman, G. (1988). Technical report on the toxicology and carcinogenesis studies of ochratoxin A in F 344/Nrats (Gavage studies). NTP Publication no. 88-2813. National Toxicology Program, US Dept of Health and Human Services.

Forgacs, J. & Carll, W. T. (1962). Mycotoxicoses. *Adv. Vet. Sci.,* **7**, 273–82.

Hartley, R. D., Nesbitt, B. F. & O'Kelly, J. (1963). Toxic metabolites of *Aspergillus flavus. Nature (London),* **198**, 1056–8.

Holzapfel, C. W., Steyn, P. S. & Purchase, I. F. H. (1966). Isolation and structure of aflatoxins M_1 and M_2. *Tetrahedron Letters,* **25**, 2799–803.

Joffe, A. Z. (1965). Toxin production in cereal fungi causing toxic alimentary aleukia in man. In *Mycotoxins in Foodstuffs,* ed. G. N. Wogan, MIT Press, Cambridge, Massachusetts, pp. 77–85.

Kinosita, R. & Shikata, T. (1965). On mouldy rice. In *Mycotoxins in Foodstuffs,* ed. G. N. Wogan, MIT Press, Cambridge, Massachusetts, pp. 111–32.

Kraus, P. V. (1978). Über die Sterigmatocystinausscheidung in Kuhmilch und über das Vorkommen von Aflatoxin in Milchprodukten. Diss. Techn. Univ. München, Federal Republic of Germany.

Krogh, P. (1977). Ochratoxins. In *Mycotoxins in Human and Animal Health,* ed. J. V. Rodricks, C. W. Hesseltine, & M. A. Mehlman. Pathotox Publishers, Park Forest South, Illinois, USA, pp. 489–98.

Krogh, P., Axelsen, N. H., Elling, F., Gyrd-Hansen, N., Hald, B., Hyldgaard-Jensen, J., Larsen, A. E., Madsen, A., Mortensen, H. P., Moller, T., Petersen, O. K., Ravnshov, U., Rostgaard, M. & Aalund, O. (1974). Experimental porcine nephropathy. Changes of renal function and structure induced by Ochratoxin A contaminated feed. *Acta Pathologica et Microbiologica Scandinavica., Sect. A. Suppl.,* **246**, 1–21.

Mayer, C. F. (1953). Endemic panmyelotoxicosis in the Russian grain belt. Part One: The clinical aspects of alimentary toxic aleukia (ATA), a comprehensive review. *Mil. Serg.,* **113**, 173–89.

Mirocha, C. J., Watson, S. & Hayes, W. (1982). Occurrence of trichothecenes in samples from South-East Asia associated with 'Yellow Rain'. In *Proc. Vth Int. IUPAC Symp. on Mycotoxins and Phycotoxins,* 1–3 September 1982, Austrian Chemical Society, Vienna, pp. 130–33.

Mirocha, C. J., Pawlosky, R. A., Chatterjee, K., Watson, S. & Hayes, W. (1983). Analysis for Fusarium toxins in various samples implicated in biological warfare in Southeast Asia. *J. Assoc. Off. Anal. Chem.,* **66**, 1485–99.

Nesbitt, B. F., O'Kelly, J., Sargeant, K. & Sheridan, A. (1962). Toxic metabolites of *Aspergillus flavus. Nature (London),* **195**, 1062–3.

Prelusky, D. B., Trenholm, H. L., Lawrence, G. A. & Scott, P. M. (1984). Non-transmission of deoxynivalenol (vomitoxin) to milk following oral administration to dairy cows. *J. Environ. Sci. Health,* **B19**, 593–609.

Robinson, T. S., Mirocha, C. J., Kurtz, H. J., Behrens, J. C., Chi, M. S., Weaver, G. A. & Nystrom, S. D. (1979). Transmission of T-2 Toxin into bovine and porcine milk. *J. Dairy Sci.,* **62**, 637–41.

Shreeve, B. J., Patterson, D. S. P. & Roberts, B. A. (1979). The carry-over of aflatoxin, ochratoxin and zearalenone from naturally contaminated feed to tissues, urine and milk of dairy cows. *Food Cosmet. Toxicol.,* **17**, 151–2.

Speijers, G. J. A., Franken, M. A. M. & Van Leeuwen, F. X. R. (1988). Subacute toxicity study of patulin in the rat; effects on the kidney and the gastrointestinal tract. *Food Chem. Tox.* **26**, 23–30.

Speijers, G. J. A. & Franken, M. A. M. (1988). Subchronic oral toxicity study of patulin in the rat. In *Food Safety and Health Protection,* ed. C. Lintas & M. A. Spandomi. Monograph Consiglio Nazionale Delle Richerche, Rome, pp. 433–6.

Stevens, A. J., Saunders, C. N., Spence, J. B. & Newnham, A. G. (1960). Investigation into 'disease' of turkey poults. *Vet. Rec.,* **72**, 627–8.

Tulasne, L. R. (1853). Mémoire sur l'ergot des Glumacées. *Ann. Sci. Nat. 3rd Ser.,* **20**, 5–56.

Van Egmond, H. P. (1989). Current situation on regulations for mycotoxins. Overview of tolerances and status of standard methods of sampling and analysis. *Food Addit. Cont.* **6**, 139–88.

Wogan, G. N. (1973). Aflatoxin carcinogenesis. In *Methods in Cancer Research,* ed. H. Bush. Academic Press, New York, pp. 309–44.

Chapter 2

Aflatoxin M₁: Occurrence, Toxicity, Regulation

H. P. van Egmond

National Institute of Public Health and Environmental Protection,
Bilthoven, The Netherlands

1. Introduction 11

2. Carry-over of Aflatoxin B₁ into Aflatoxin M₁ in Milk 12
    2.1 Conversion at high aflatoxin levels
    2.2 Conversion at low aflatoxin levels
        2.3 Conclusion

3. Occurrence of Aflatoxin M₁ in Milk and Milk Products 16
    3.1 Surveys in the 1970s
    3.2 Surveys in the 1980s
        3.3 Conclusion

4. Toxic Effects of Aflatoxin M₁ 23
    4.1 Availability of aflatoxin M₁ for animal studies
    4.2 Short and medium term toxicity studies
        4.3 Carcinogenicity studies
        4.4 Conclusion

5. Regulation 31
        5.1 Introduction
    5.2 Aflatoxin regulation for animal feedstuffs
    5.3 Aflatoxin M₁ regulation for dairy products
        5.4 Reference materials
        5.5 Conclusion

6. Summary 47

1 INTRODUCTION

Shortly after the discovery of the aflatoxins, Allcroft & Carnaghan (1963) suggested that aflatoxin residues might occur in milk and other animal products from animals that had ingested aflatoxins with the feedstuff. This suggestion was of major importance because of the frequent and ubiquitous use of groundnut meal in animal feedstuffs

11

and the frequent presence of aflatoxins in that groundnut meal. A toxic factor in the milk of cows fed with aflatoxin B_1-contaminated feedstuff caused the same toxic effects in young ducklings as aflatoxin B_1. When curdling the milk, the toxic factor appeared in the casein fraction. The nature of the 'milk toxin' was further studied by De Iongh *et al.* (1964). They showed by thin layer chromatography on silica gel that the toxic factor had a blue fluorescence as aflatoxin B_1, but it had an R_f value much lower than that of aflatoxin B_1. The 'milk toxin' was also found in the liver, kidney and urine in sheep that had been administered a single dose of mixed aflatoxins B_1, B_2, G_1 and G_2. A trivial name, aflatoxin M, was suggested to indicate its original isolation from milk.

The structure of aflatoxin M was elucidated by Holzapfel *et al.* (1966) who found in fact two components that could be separated by paper chromatography. They were designated aflatoxins M_1 and M_2. These metabolites seemed to be the 4-hydroxy derivatives of aflatoxins B_1 and B_2 respectively. The subscripts of aflatoxins M_1 and M_2 related to the R_f values, as is the case with the aflatoxins of the B- and G-group. The chemical structures of the aflatoxins M_1 and M_2 are given in Fig. 1.

Because aflatoxin B_1 (AFB_1) is the most important of the aflatoxins, considered from the viewpoints of both toxicity and occurrence, most attention was given to its metabolite aflatoxin M_1 (AFM_1). Recently, a new hydroxy derivative of aflatoxin B_1 has been detected in milk (Lafont *et al.*, 1986*a*). This metabolite was named aflatoxin M_4 (see Fig. 1). Aflatoxin M_4 seems to co-occur in certain milks next to AFM_1, in amounts up to *c.* 16% of the AFM_1 content (Lafont *et al.*, 1986*b*). Aflatoxin M_4 is also produced *in vitro* by certain strains of *Aspergillus parasiticus*. Tests have indicated that aflatoxin M_4 is more toxic and carcinogenic than aflatoxins B_1 and M_1 (Lafont & Lafont, 1987). Because the knowledge about aflatoxin M_4 is at present very limited, this toxin will not be further discussed here.

2 CARRY-OVER OF AFLATOXIN B_1 INTO AFLATOXIN M_1 IN MILK

Since the discovery that dairy cows consuming rations contaminated with AFB_1 excrete AFM_1 in their milk, various studies have been undertaken to establish the carry-over rates.

2.1 Conversion at High Aflatoxin Levels

Probably Van der Linde *et al.* (1964) were the first to study the quantitative transmission of AFB_1 into the milk of dairy cows. They

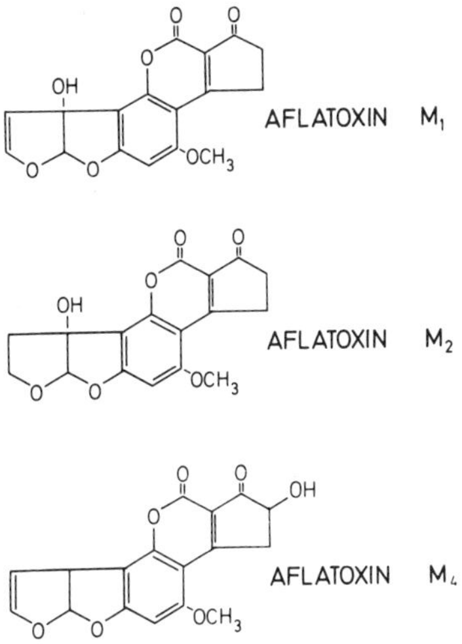

Fig. 1. Chemical structures of aflatoxins M_1, M_2, and M_4.

used two cows with high milk yields (28 litres per day) and two cows with low milk yields (12 litres per day). Each of the cows was given a daily ration of 2 kg of groundnut meal contaminated with AFB_1 at a level of 4 mg/kg, for a period of 18 days. The milk of the cows was monitored daily for AFM_1 by a chemical test (thin layer chromatography) and by a bioassay (one-day-old ducklings). The toxin could be readily detected in the milk 12–24 h after the first AFB_1 ingestion. After a few days the toxin content in the milk reached a high value. The amount of toxin in the milk was less than 1% of the ingested AFB_1. After the intake of AFB_1 had ended, the AFM_1 concentration in the milk dropped down to an undetectable level after three days. There appeared to be no relationship between the toxin content in the milk and the daily milk yield.

The study of Van der Linde *et al.* (1964) was followed by various other studies on the carry-over of AFB_1. Reviews are given by Rodricks & Stoloff (1977) and Sieber & Blanc (1978). The reviewed data given by Rodricks & Stoloff (1977) showed that the ratios of the concentration of AFB_1 in the cattle feed to that of AFM_1 in the milk ranged from 34–1600 with an average ratio near 300. Sieber & Blanc (1978) expressed the results in a different way. They calculated the excreted amount of AFM_1 as a percentage of AFB_1 and found estimates ranging from 0–4% with an average ratio of *c.* 1%. Kiermeier *et al.* (1975) indicated that concentrations of AFM_1 in milk may vary largely from animal to animal, even

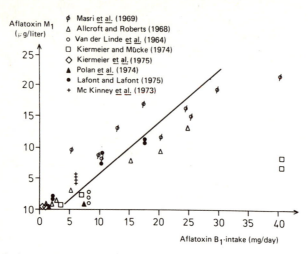

Fig. 2. Relationship between aflatoxin B_1 intake with the feedstuff and aflatoxin M_1 secretion in the milk. (After Sieber & Blanc, 1978; courtesy of *Mitt. Gebiete Lebensm. Hyg.*)

for animals from the same herd. Mertens (1979) attributed the differences to the experimental techniques involved. While some researchers have measured the total excretion of a single dose of aflatoxin, the method most relevant to estimation of milk contamination in the field is a steady output of aflatoxin per day. The data of several researchers (Van der Linde *et al.*, 1964; Masri *et al.*, 1969; Kiermeier, 1973*a*; McKinney *et al.*, 1973; Polan *et al.*, 1974) suggest that it takes 3–6 days of constant daily ingestion of AFB_1 before steady state excretion of AFM_1 in the milk is achieved, whereas AFM_1 becomes undectetable 2–4 days after withdrawal from the contaminated diet.

Sieber & Blanc (1978) presented the conversion data of several studies graphically (Fig. 2). They concluded there was a linear regression between the AFB_1 intake with the feedstuff and the AFM_1 secretion in the milk. The corresponding regression equation was $y = -2·55 + 0·84x$ ($r^2 = 0·73$; $n = 43$).

2.2 Conversion at Low Aflatoxin Levels

Most of the studies until 1980 were carried out with feedstuffs containing dietary levels of AFB_1 that were often much higher than the existing official limits for aflatoxins in the feed for dairy cattle. Usual legal limits for AFB_1 in dairy rations are 10–20 μg/kg (Van Egmond, 1989). Figure 2 and the regression equation of Sieber & Blanc (1978) are not applicable for these (more realistic) levels of contamination. Little detailed information had been published on the carry-over of aflatoxins at these low dietary levels until Patterson *et al.* (1980) and Lafont *et al.* (1980) undertook such studies.

In the study of Patterson *et al.* (1980), a group of six dairy cows were given a daily diet contaminated with approximately 10 μg/kg of AFB_1. During a period of seven days, the milk obtained from the herd was analysed for AFM_1. The concentration of AFM_1 was found to vary between 0·01 and 0·33 μg/litre with a mean value of 0·19 μg/litre. An average of approximately 2·2% of ingested AFB_1 appeared in the milk daily as AFM_1. As published previously by Kiermeier *et al.* (1977), concentrations of AFM_1 in milk appeared to vary from animal to animal, from day to day and from one milking to the next.

In contrast with the experimental results of Patterson *et al.* (1980) are those of Lafont *et al.* (1980). They conducted an experiment with 16 cows, some in the early and the rest in the late lactation period. The cows were divided into four groups, each group receiving AFB_1 at 0·09, 0·18, 0·86 and 2·58 mg/day/animal, respectively, as compared to 0·15–0·23 mg/ day/animal in the study of Patterson *et al.* (1980). During a period of 12 days the AFM_1 content in the milk of the various groups of cows was determined. Lafont *et al.* (1980) concluded that the quantity of secreted AFM_1, expressed as a percentage of the parent aflatoxin, varied from 0·14–0·34% (average 0·22%) in the animals in the late lactation period, and from 0·66–0·95% (average 0·78%) in the cows producing some 20 litres of milk per day. The daily milk yield of the latter group was comparable to the daily milk yield of the cows in the experiment of Patterson *et al.* (1980), but the experimentally estimated carry-over rates differed by a factor of three. Moreover, Lafont *et al.* (1980) were unable to determine any AFM_1 in the milk of the cows with a daily AFB_1 intake of 0·18 mg as compared to 0·19 mg daily AFB_1 intake of the cows in the experiment of Patterson *et al.* (1980). Lafont *et al.* (1980) did not indicate the limit of detection of their analytical method, but from their results it can be derived that this limit was low enough to detect a carry-over rate of 1% at the 0·18 mg AFB_1 intake level.

The differences in results of the studies of Lafont *et al.* (1980) and Patterson *et al.* (1980) make it difficult to conclude what the conversion rate is at low levels of AFB_1 in feedstuffs for dairy cattle. The estimated percentages of 0·78% and 2·22% fall within the range of 0–4% carry-over as estimated by Sieber & Blanc (1978). On the basis of a daily consumption per cow of 6 kg of feedstuff contaminated with AFB_1 at 10 µg/kg, and a milk production of 20 litres a day, an AFM_1 level in the milk ranging from 0·02–0·07 µg/litre could be expected. This range is in the order of the (proposed) acceptable levels in various European countries (see Section 5.3.2). More or less feed consumption will lead to a proportional increase or decrease respectively of the expected AFM_1 level in the milk.

2.3 Conclusion

The ingestion of AFB_1 in feedstuffs by the dairy cow leads to the excretion of AFM_1 in the milk. Experimentally established carry-over rates for high and low contaminated rations average 1–2%, the percentages varying from animal to animal, from day to day and from one milking to the next.

3 OCCURRENCE OF AFLATOXIN M_1 IN MILK AND MILK PRODUCTS

AFM_1 enters the human food by indirect contamination. AFM_1 may occur in animal organs and tissues, e.g. kidneys, and in animal products, e.g. milk and other dairy products (see Chapter 5) after consumption of AFB_1-contaminated feeds by the animal. The sources of aflatoxin contamination in animal feedstuffs may vary geographically. Many feedstuff ingredients may contain aflatoxins, with groundnut meal, cottonseed meal and maize meal among the most important. The contamination of agricultural crops with aflatoxins is not restricted to the developing countries, where both climatic and technological circumstances stimulate aflatoxin formation. Even technologically well-developed countries such as the USA suffer from large-scale outbreaks of AFB_1 from time to time (Applebaum *et al.*, 1982).

To control AFM_1 in milk and milk products, several countries have established specific regulations for aflatoxins in animal feedstuffs for dairy cattle and for AFM_1 in milk (see Sections 5.2 and 5.3). Surveillance programmes for aflatoxins in dairy cattle feed and for AFM_1 in milk (products) are carried out in many countries. Surveys on aflatoxins in dairy feedstuffs will not be discussed here. Data on the occurrence of AFM_1 in milk (products) are given hereafter.

3.1 Surveys in the 1970s

The results of various surveys of milk for AFM_1 carried out in various countries in the late 1960s and the 1970s have been summarized by Brown (1982) (see Table 1). A seasonal trend in milk contamination was noted in a few of these surveys, with lower AFM_1 levels in milk in the summer months. This phenomenon was attributed to the fact that the cows are receiving less concentrated feeds in summertime, when they are grazing. In all surveys, positive samples were found (except in the summer survey of the German Democratic Republic), and there were samples with AFM_1 levels exceeding $0.05 \mu g/kg$. In various studies, samples were reported with levels $> 0.5 \mu g/kg$ ($0.05–0.5 \mu g/kg$ is the current range of (proposed) tolerance values for AFM in milk, with the exception of infant milk, for which lower levels exist — see Section 5.3.2). It is amazing to notice the relatively high levels of AFM_1 in milks in India and the German Democratic Republic, taking account of the fact that the amount of AFM_1 in the milk is approximately 1% of the amount of AFB_1 in the feedstuff (see Section 2).

3.2 Surveys in the 1980s

The establishment of new regulations in the late 1970s and in the 1980s to control the aflatoxin contents of dairy rations (see Section 5.2) led to the expectation that AFM_1 levels in milk would reduce in the 1980s. At the same time, newer analytical developments would make detection possible at lower concentrations of AFM_1.

To get an overview of worldwide data on AFM_1 occurrence in milk and milk products in the 1980s, the author of this chapter held an enquiry in 1986 among a number of countries expected to occasionally or routinely carry out analyses for AFM_1. Twelve countries sent in data that were relevant to the situation in the 1980s. The results of surveys of milk and milk powder analysed for AFM_1 in the 1980s are summarized in Table 2.

Compared to the studies in the 1970s the situation for AFM_1 contamination of milk and milk-powder seems to have improved. In general, both incidences of positive samples and AFM_1 levels were

H. P. van Egmond

TABLE 1

Occurrence of aflatoxin M_1 in milk; surveys carried out in the late 1960s and in the 1970s (Brown, 1982)

Country	No. positive/ no. samples	Proportion positive[a] (%)	Range of concentrations (µg/kg)	Reference
South Africa	5/21	24	0·02–0·2	Rodricks & Stoloff (1977)
Federal Republic of Germany	28/61	46	0·04–0·25	Kiermeier (1973b)
Federal Republic of Germany	118/260	45	0·05–0·30	Polzhofer (1977)
Federal Republic of Germany	79/419	19	0·05–0·54	Kiermeier et al. (1977)
USA	191/302	63	trace–>0·7	Stoloff (1980)
Belgium	42/68	62	0·02–0·2	Van Pée et al. (1977)
India	3/21	14	up to 13·3	Paul et al. (1976)
German Democratic Republic (winter)	4/36	11	1·7–6·5	Fritz et al. (1977)
German Democratic Republic (summer)	0/12	0	<0·1	Fritz et al. (1977)
Netherlands (winter)	74/95	82	0·03–0·5	Schuller et al. (1977)
UK	85/278	31	0·03–0·52	Ministry of Agriculture, Fisheries and Food (1980)

[a]N.B. The percentages given should not be directly compared with each other because they also depend, of course, on the limits of detection of the methods of analysis used.

TABLE 2

Occurrence of aflatoxin M_1 in liquid and powdered milk; selected surveys in the 1980s

Country	Period	Type of samples	No. positives/ no. samples	Proportion positive[a] (%)	Range of concentrations[b] (µg/kg)	Reference	Remarks
Austria	February 1983	Milk-powder	0/1	0	<0·03	Brandl (1986)	
	November 1983	Milk-powder	0/20	0	<0·03		
	October 1984	Milk-powder	0/39	0	<0·01		
	May 1985	Milk-powder	0/2	0	<0·01		
	April 1986	Milk-powder	0/3	0	<0·01		
Belgium	Winter 1980-1	(Individual farm milk) Milk-powder	3/25	12	0·01	Srebrnik (1981)	
		Milk	2/32	6	0·09-0·15		
	Summer 1981	(Commercial blended milk) Infant milk	0/15	0	<0·01	Srebrnik (1981)	
		Milk-powder	1/30	3	0·05		
		Evaporated milk	0/7	0	<0·03		
		Milk	0/13	0	<0·03		
	Winter 1981-2	(Commercial blended milk) Milk	11/56	20	0·03-0·10	Srebrnik (1981, 1982)	
		Milk-powder	7/49	14	0·01-0·10		
		Evaporated milk	1/22	5	0·03-0·10		
		Infant milk	5/23	22	0·03-0·14		
	Winter 1982-3	(Commercial blended milk) Milk	7/71	10	0·02-0·08	Srebrnik (1982)	
		Evaporated milk	0/2	0	<0·03		
		Infant milk	5/20	25	0·01-0·02		
	Winter 1983-4	Commercial blended milk	64/124	52	0·02-0·5		
	Winter 1984-5	Commercial blended milk	29/233	12	0·02-0·15	Srebrnik (1986)	
	Winter 1985-6	Commercial blended milk	0/89	0	<0·03		

(continued)

TABLE 2—*contd.*

Country	Period	Type of samples	No. positives/ no. samples	Proportion positive[a] (%)	Range of concentrations[b] (µg/kg)	Reference	Remarks
China (North)	1981	Milk-powder	0/82	0	<0·05	Kong Zhongfu *et al.* (1985)	
(South)		Milk-powder	61/85	72	0·05–0·5		
(Guangxi)	April–June 1983	Milk	17/17	100	0·1–0·3	Hu Zhuohan *et al.* (1985)	
(Shanghai)		Milk-powder	50/54	93	0·011–0·167		
		Milk	45/81	56	0·02–0·13		
China (Taiwan)	March–September 1986	Fresh milk, local produce	0/56	0	<0·1	Yau-Huei Wei (1986)	
		Milk-powder, import	0/161	0	<0·1		
Finland	June–October 1982	Milk-powder	0/3	0	<0·1	Hintikka (1986)	
	1986	Milk-powder	0/14	0	<0·1	Björk (1986)	
France	January–March 1981	Milk	313/394	79	0·05–>0·5	Carbonel & Sudreau (1981)	
	November–December 1981	Milk	14/169	8	0·05–<0·5		
	January–May 1982	Milk	0/72	0	<0·05	Frémy *et al.* (1982b)	
		Milk-powder	34/238	14	0·05–<0·5		
	January–December 1982	Milk	90/243	37	0·02–<0·3	Carbonel & Sudreau (1984)	
	January–December 1983	Milk	78/474	16	0·02–<0·3		
	January–March 1983	Milk	10/238	4	0·05–<0·5		
	May–July 1983	Milk	1/209	0·5	0·05–<0·5	Frémy *et al.* (1983)	
	September–November 1983	Milk	0/245	0	<0·05		
	January–November 1984	Milk	33/341	10	0·02–<0·3	Carbonel & Sudreau (1984)	
	January–March 1984	Milk	0/202	0	<0·05		
	September–November 1984	Milk	1/205	0·5	0·05–<0·5	Frémy *et al.* (1984)	

Country	Date	Product	Positive/total	%	Range	Reference	Comment
Germany (Federal Republic)	January–March 1985	Milk	6/289	2	0·05–<0·5	Frémy et al. (1985)	
	September–November 1985	Milk	0/215	0	<0·05		
	November 1984–April 1985	Herds' bulk milk	/135ᶜ		0·003–0·080		
		Pasteurized commercial milk	/132		0·003–0·060		
		Milk-powder	/46		0·003–0·020		
				Not given		Heeschen (1986)	Samples collected in the county of Schleswig-Holstein
	November 1985–April 1986	Herds' bulk milk	/242		0·003–0·100		
		Pasteurized commercial milk	/16		0·007–0·013		
		Milk-powder	/33		0·005–0·034		
Ireland	May–June 1986	Herds' bulk milk	/74		0·003–0·020	Walsh (1986)	Samples collected in various parts of the country
	November 1981	Skim milk-powder	0/5	0	<0·015		
	December 1981	Skim milk-powder	0/7	0	<0·015		
	January 1982	Skim milk-powder	0/7	0	<0·015		
	February 1982	Skim milk-powder	0/11	0	<0·015		
	March 1982	Skim milk-powder	0/3	0	<0·015		
	April 1982	Skim milk-powder	0/3	0	<0·015		
Italy	1982	Milk-powder	64/279	23	0·002–0·010	Riberzani et al. (1983)	
	Spring 1983–4	Raw whole milk	8/31	26	0·005–0·091	Visconti et al. (1985)	Samples collected in Southern Italy
		Heat-treated milk	59/66	89	0·004–0·150		
		Milk-powder	9/9	100	0·001–0·028		
	Spring 1984	Raw whole milk	27/52	52	0·005–0·146	Boccia et al. (1985)	Samples collected in Central Italy
	Spring 1984	Commercial milk	12/18	67	0·005–0·030	Piva et al. (1985)	Samples collected in Northern Italy
		Raw whole milk	34/82	41	Not given		
The Netherlands	January–April 1985	Bulk-milk and milk-powder	169/209	81	0·01–0·07	Agricultural Advisory Committee (1986)	
	May–August 1985	Bulk-milk and milk-powder	153/207	74	0·01–0·05		
	September–December 1985	Bulk-milk and milk-powder	188/207	91	0·01–0·09		

(continued)

TABLE 2—contd.

Country	Period	Type of samples	No. positives/ no. samples	Proportion positive[a] (%)	Range of concentrations[b] ($\mu g/kg$)	Reference	Remarks
	January–April 1986	Bulk-milk and milk-powder	174/212	82	0·01–0·09	Agricultural Advisory Committee (1987)	
	May–August 1986	Bulk-milk and milk-powder	115/204	56	0·01–0·07		
	September–November 1986	Bulk-milk and milk-powder	165/202	82	0·01–0·05		
Spain	February–May 1983	Commercial milk (raw, sterilized, pasteurized, concentrated)	7/95	7	0·020–0·040	Burdaspal et al. (1983)	Samples collected in Madrid
Sweden	April–May 1983	Milk	0/60	0	<0·050	Möller & Andersson (1983)	
	April 1983	Milk-powder	0/24	0	<0·050		
	April 1983	Milk	13/13	100	0·005–0·036	Borgström (1983)	
	December 1985	Milk-powder	14/14	100	0·006–0·057		
	January–March 1986	Milk	239/268	90	0·005–0·312	Möller (1987)	
	April–June 1986	Milk	118/271	44	0·005–0·050		
United Kingdom	1981–3 (monthly samples)	Milk-powder	35/277	13	0·01–0·4	Gilbert et al. (1984)	
		Milk, collected at farm gate	24/409	6	0·01–0·78		

[a] N.B. The percentages given should not be directly compared with each other because they also depend, of course, on the limit of detection of the methods of analysis used.
[b] For milk-powder calculated on the basis of reconstituted milk (dilution factor: 10×). Ranges are given for positive samples. If no positive samples were found, the limit of detection is given, where known. N.B. Although mean and median concentration were made available for some countries, they were left out for reasons of uniformity.
[c] No. of samples given only.

lower in the 1980s than in the 1970s. It must be emphasized, however, that only a limited number of countries was involved in both the surveys of the 1970s and the 1980s, which makes it difficult to draw firm conclusions. The noted decrease in AFM_1 concentrations is probably the effect of legislative action on feedstuff contamination with aflatoxin(s), which has come into force in many countries. However, some countries currently (1989) apply limits of 0·01 μg/kg for AFM_1 in milk for infants and infant foods (see Section 5.3.2) and it is obvious that many countries have problems in meeting these very low requirements. The summarized data also reflect the improvements in the limits of detection of the methodology used. With 0·02 μg/kg obviously being the lowest detectable amount in the 1970s, AFM_1 levels down to 0·003 μg/kg were reported in the 1980s!

Although some countries reported data on AFM_1 in cheese, they are rather scarce, and therefore they have not been tabulated. In general AFM_1 was not detectable or occurring in cheese in concentrations lower than the current legal limits (0·2–0·25 μg/kg) that exist for AFM_1 in cheese in a few countries.

Not included in the considerations above nor in Table 2, are the very recent data of monitoring programmes carried out in the USA and Europe in late 1988 and early 1989. These data show a significant increase of AFM_1 levels in milk (products) as a result of the feeding to dairy cattle with US maize products, contaminated with aflatoxin B_1. The contamination was due to the severe drought that occurred in the US Midwest in summer 1988, and that created hospitable conditions for aflatoxin formation by *Aspergillus flavus*. This recent aflatoxin outbreak and its serious impact for the contamination of dairy products has caused concern in several countries.

3.3 Conclusion

Results of surveys of milk for AFM_1 in the 1970s and those of surveys of milk, milk-powder and cheese in the 1980s have shown that the AFM_1 incidences and levels have considerably decreased in recent years, although an increase occurred again in 1989. Most of the samples of milks, milk-powders and cheeses taken in the 1980s in various countries have AFM_1 levels at ranges that are acceptable for these products. However, for milk and milk-powder intended to be used for infants and infant formulae, some countries adopt very low AFM_1 tolerances, which will not be easily enforceable.

4 TOXIC EFFECTS OF AFLATOXIN M₁

4.1 Availability of Aflatoxin M₁ for Animal Studies

Since the discovery of AFM_1 there have been several studies on the toxic effects of the milk toxin to laboratory animals. In comparison with AFB_1, however, relatively little is known about the toxicity of AFM_1. This is mainly because insufficient quantities of the pure compound have precluded extensive toxicity experiments to be carried out.

There are, in principle, several possibilities of obtaining AFM_1:

(1) Extraction from contaminated milk or urine. Because the AFM_1 concentration is very low and the purification procedure tedious, this approach is not feasible.

(2) Chemical conversion of AFB_1 into AFM_1. A four step synthesis, described by Christon *et al.* (1985) gave an overall yield of pure AFM_1 from AFB_1 of only 0·03%. Because large amounts of the highly carcinogenic AFB_1 would need to be handled to obtain small amounts of AFM_1, this approach would not be attractive either.

(3) Total synthesis of AFM_1. Büchi & Weinreb (1971) succeeded in preparing 50 mg of racemic AFM_1, which was used for biological studies (Pong & Wogan, 1971; Wogan & Paglialunga, 1974). The investigators reported, however, that two of the intermediates were unstable and difficult to handle, particularly on scales larger than those described. The synthesis also suffered from a low overall yield and therefore a new synthesis was developed (Büchi *et al.*, 1981). Attempts at the State University of Utrecht, The Netherlands, in cooperation with the National Institute of Public Health and Environmental Protection, The Netherlands, to synthesize gram amounts of AFM_1 according to the routes as described by Büchi & Weinreb (1971), and Büchi *et al.* (1981) failed, due to serious scaling up problems (Van Egmond & Stavenuiter, 1985). Another disadvantage of the synthesis approach is the fact that a racemic mixture of AFM_1 is formed whereas in nature only one enantiomer occurs.

(4) Biotransformation of AFB_1 to AFM_1 using induced liver enzymes. This way of production of AFM_1 yielded sufficient quantities for metabolism studies (Rice & Hsieh, 1982), but not for chronic toxicity studies.

(5) Biosynthesis of AFM_1 by *Aspergillus flavus* or *Aspergillus parasiticus*.

Wiley & Waiss (1968) described a method in which *A. flavus* is cultured on rice. In addition to substantial amounts of aflatoxins B_1 and B_2 (and sometimes G_1 and G_2), some strains of *A. flavus* and *A. parasiticus* produce minute amounts of aflatoxins M_1 and M_2 (see also Chapter 3). The rice culturing procedure was also described by Stubblefield *et al.* (1970), who isolated the material for analytical purposes and by Hsieh *et al.* (1986*a*) who isolated AFM_1 for toxicological studies. Although the amounts of AFM_1 obtained in this fermentation method are only in the mg range, this method is, as yet, the most feasible of the mentioned possibilities of producing AFM_1.

4.2 Short and Medium Term Toxicity Studies

Due to the limited supply of pure AFM_1, most toxicity assays for AFM_1 were done for short and medium terms of exposure.

4.2.1 Ducklings
The first detection methods for aflatoxins were bioassays with newly hatched ducklings (Allcroft & Carnaghan, 1963). These animals are extremely sensitive for both AFB_1 and AFM_1 and LD_{50} values of 12–16 μg per animal were reported (Holzapfel *et al.*, 1966; Purchase, 1967). Histopathological examinations revealed liver lesions and kidney damage. The hepatic ultrastructural changes found after dosing with AFM_1 were qualitatively similar to that produced by AFB_1. Necrosis of the kidney tubuli which was observed for AFM_1, but not for AFB_1, was explained by the fact that AFM_1 is more polar through the presence of a hydroxy group in the molecule. Therefore it is better soluble in aqueous solvents and easier excreted with the urine.

Frémy *et al.* (1982*a*) compared the effects of oral administration to newly hatched ducklings of milks naturally and artificially contaminated with AFM_1 at a level of 14 μg/kg. Five ml amounts of milk were given five times a day during four consecutive days. The investigators found that the naturally-contaminated milk produced less lesions than the artificially-contaminated milk. This would suggest differences in the bio-availability of naturally- and artificially-occurring AFM_1. In a medium term toxicity study Ferrando *et al.* (1984) incorporated milk-powder naturally contaminated at a level of 25 μg/kg into the diet of newly hatched ducklings. At the dietary percentage of 30% milk-powder no toxic effects were noticed during the trial period which lasted 69 days.

4.2.2 Rats

In addition to the one-day-old duckling studies, there have been some investigations on the toxic effects of synthetically-produced (racemic) AFM_1 in Fisher rats (Pong & Wogan, 1971). Synthetic and natural AFB_1 were also administered in the same study. The synthetic AFB_1 and AFM_1 were produced by Büchi *et al.* (1967) and Büchi & Weinreb (1971) respectively. Both aflatoxins consisted of racemic mixtures of two enantiomers. It was shown that both synthetic aflatoxins were lethal to male rats at a single dose of 1·5 mg/kg body weight, whereas natural (non-racemic) AFB_1 was lethal at a dose of 0·6 mg/kg. Changes in the liver parenchymal cells such as dissociation of ribosomes from the rough endoplasmic reticulum, and proliferation of the smooth endoplasmic reticulum were observed. Qualitatively these changes were similar for the racemic and natural toxins. Quantitatively the structural changes induced by the synthetic, racemic toxins at 1·0 mg/kg were indistinguishable from those caused by natural AFB_1 at a dose of 0·5 mg/kg. The data suggested that AFM_1 acts through the same mechanism as AFB_1 in causing acute toxicity and subcellular alterations, and that only the natural occurring isomer of each aflatoxin is biologically active.

4.3 Carcinogenicity Studies

After the finding that AFB_1 is a potent hepatocarcinogen in the rat (Butler & Barnes, 1968) and in all other species of laboratory animals tested (Wogan, 1973), serious concern arose about the possible carcinogenicity of AFM_1 because of its structural and toxicological similarity to AFB_1. The lack of sufficient quantities of pure natural AFM_1 initially precluded extensive carcinogenicity studies with mammals. Therefore the first carcinogenicity studies with natural AFM_1 were carried out with rainbow trout, a species very sensitive to AFB_1 (Sinnhuber *et al.*, 1974; Canton *et al.*, 1975). Later, genotoxicity studies (Lutz, 1978, 1979; Green *et al.*, 1982) were undertaken. One carcinogenicity study with synthetically–produced AFM_1 with rats was described (Wogan & Paglialunga, 1974). More recently, when Hsieh *et al.* (1986*a*) succeeded in preparing reasonable amounts of AFM_1 from cultures of *A. flavus* on rice, limited rat studies were undertaken with natural AFM_1 (Cullen *et al.*, 1987).

4.3.1 Rainbow trout
Sinnhuber *et al.* (1974) conducted a study in which duplicate lots of

rainbow trout received experimental diets containing AFM_1 at 0, 4, 16, 32 and 64 μg/kg feed. Control diets with AFB_1 at 4 μg/kg and with cyclopropenoid fatty acids at 100 mg/kg feed were added for comparison. Cyclopropenoid fatty acids were known to promote the carcinogenicity of AFB_1 (Sinnhuber *et al.*, 1968). The fish were fed the experimental diets continuously for 12 months, after which they were kept on the control diet. Random samples were taken at 4, 8, 12 and 16 months. Certain groups were held for 20 months to determine the effect of maturation on tumour development. In addition, the effect of limited oral intake of AFM_1 was determined by feeding groups of trout for 5, 10, 20 and 30 days at an AFM_1 level of 20 μg/kg, with and without cyclopropenoid fatty acids. Levels of AFM_1 at 4 and 16 μg/kg and AFB_1 at 4 μg/kg produced 13%, 60% and 48% incidences respectively of hepatoma in 12 months. Dietary levels of AFM_1 at 4 μg/kg with cyclopropenoid fatty acids gave a 70% incidence of liver cancer in 8 months, so that a promoter effect was evident, similar as for AFB_1. A significant number of mortalities occurred among female trout having AFM_1-induced hepatomas at the time of maturation (16–20 months) in contrast to no mortality in males. Trout that received AFM_1 at levels of 20 μg/kg during 5–30 days developed a 3–12% incidence of hepatoma in 12 months. The investigators concluded that AFM_1 is a potent liver carcinogen, but less potent than AFB_1.

Canton *et al.* (1975) conducted a similar study with groups of rainbow trout fed diets containing AFM_1 at 0, 5·9 and 27·3 μg/kg and AFB_1 at 5·8 μg/kg respectively for 16 months. Necropsy of fish killed after 5, 9 and 12 months revealed ceroid degeneration of the liver in all three groups and in the control group, but no tumours or preneoplastic changes. Autopsy of survivors at month 16, however, revealed 13% with hepatocellular carcinoma and 23% with hyperplastic nodules in the group fed AFB_1 at 5·8 μg/kg, and 2% with hepatocellular carcinoma and 6% with hyperplastic nodules in the group fed AFM_1 at 27·3 μg/kg. No evidence of such lesions was seen in the group fed AFM_1 at 5·9 μg/kg and in the control group. Comparing their results with those of the study of Sinnhuber *et al.* (1974), the investigators concluded that the different trout strains used may have been responsible for the differences in results. They confirmed the finding of Sinnhuber *et al.* (1974) that AFM_1 is less carcinogenic than AFB_1.

4.3.2 Genotoxicity and mutagenicity tests
Lutz (1978, 1979) applied a short-term test to measure the binding of AFM_1 to liver DNA in the rat. In this in-vivo binding test a certain dose

of radioactive chemical is administered and after a waiting period of a few hours, liver DNA is isolated and the radioactivity on the DNA is measured. From the radioactivity on the DNA, a covalent binding index (CBI) can be calculated. The CBI is a measure for the damage of DNA resulting from the exposure of a certain dose of chemical. The CBI correlates to the carcinogenicity of the compound because if damaged DNA is not repaired before a replication occurs, it can lead to mutations which may transform the cell and give rise, in the end, to a chemically induced tumour (Duncan *et al.*, 1969). Lutz (1978) found a binding ratio of *c.* 2×10^{-6} at 6 h after administration, a value 15 times below the binding ratio of AFB_1. Despite this difference with AFB_1, the outcome of the experiment suggested that AFM_1 still belongs to the group of potent hepatocarcinogens.

Wong & Hsieh (1976) studied the potency of several aflatoxin metabolites, including AFM_1, in the Ames mutagenicity test. In this test procedure the mutagenicity of chemicals is measured with the help of a strain of *Salmonella typhimurium* (Ames *et al.*, 1975). It was found that AFM_1 had 3·2% of the mutagenic activity of AFB_1. Green *et al.* (1982) determined genotoxicity of AFM_1 by an assay for unscheduled DNA repair. This assay uses primary cultures of rat hepatocytes and it is based on the ability of the hepatocytes to metabolize chemicals to forms that react with DNA and then respond to the damage by repairing the DNA (Williams, 1977). With few exceptions there is a good correlation between the carcinogenicity of a chemical and a positive response in the DNA repair assay (Sirica & Pitot, 1979). The experiments showed a lowest genotoxic dose of AFM_1 of 0·05 µg/culture, whereas the lowest genotoxic dose of the more potent AFB_1 was 0·025 µg/culture.

4.3.3 Rats

The availability of milligram quantities of synthetically-produced (racemic) AFM_1 enabled Wogan & Paglialunga (1974) to conduct a limited carcinogenicity study. At the start of the experiment, weanling Fisher rats (30 males) were dosed 25 µg synthetic AFM_1 per day by intubation, five days a week for eight consecutive weeks. For comparison, a group of rats were given natural AFB_1 at the same concentration and under similar conditions as for AFM_1. A control group which did not receive any aflatoxin was also included in the study. The experiment was terminated when animals died or began to show clinical deterioration. Only one rat (3%) dosed with AFM_1 developed a hepatocellular carcinoma, whereas 28% had liver lesions diagnosed as early or advanced preneoplastic lesions. All rats treated with AFB_1

developed tumours, whereas the controls showed no significant liver pathology. It was concluded that AFM_1 had a much lower carcinogenic potency than AFB_1, even though the effective dose of AFM_1 was probably only half that of AFB_1 (assuming only one isomer of the racemic mixture to be biologically active).

Cullen *et al.* (1985, 1987) described the hitherto only published carcinogenicity study with natural AFM_1 in rats. Male Fisher rats were chosen as the animal model in view of their high sensitivity to AFB_1 carcinogenicity (Wogan, 1973). Four test groups of 62 animals were maintained on diets containing natural AFM_1 at 0, 0·5, 5 and 50 μg/kg. A positive control group maintained on a diet containing AFB_1 at 50 μg/kg was included. The feeding experiment started when the animals were seven weeks of age and was continued until the animals were sacrificed between 18 and 22 months of age. Hepatocellular carcinoma were detected in 5% of the rats and neoplastic nodules were found in 16% of the rats fed AFM_1 at 50 μg/kg, between 19 and 21 months. No nodules or carcinomas were observed in the lower AFM_1 dose groups. Ninety-five per cent of the rats fed the diet containing AFB_1 at 50 μg/kg developed hepatocellular carcinomas. In the 50 μg/kg AFM_1 group three rats also developed intestinal carcinomas, while no intestinal neoplasms were observed in any of the other groups. The authors suggested that the greater polarity of AFM_1 compared to AFB_1 might lead to poor absorption from the digestive tract and that this might be associated with the higher incidence of intestinal tumours. It was concluded that AFM_1 is a hepatic carcinogen, although with a potency of about 2–10% of that of AFB_1, and that AFM_1 is also an intestinal carcinogen.

4.3.4 Risk assessment

The frequent detection of AFM_1 in commercial milk samples and dairy products (see Section 3), the high consumption of these products, especially in infant populations and the probable carcinogenicity of AFM_1 has led to incidental assessment of cancer risk due to AFM_1 in the USA (Hsieh, 1985; Hsieh *et al.*, 1986*b*). They calculated a worst case of daily intake of AFM_1 in the USA of 218 ng/*capita*/day, or 4·4 ng/kg body weight/day. Assuming a *per capita* food consumption rate of 2 kg of moist food per day, an average dietary concentration of approximately 0·11 μg/kg was estimated. At this level, no carcinomas in the Fisher rat were observed (Cullen *et al.*, 1987). The authors considered humans to be more resistant than the rat to the hepatocarcinogenicity of AFM_1, based on three observations with AFB_1. First, rhesus monkeys are much more resistant than the rat to the hepatocarcinogenicity of AFB_1 (Seiber *et al.*, 1979), and the postmitochondrial preparation of the human liver metabolized AFB_1 in a manner very similar to that of the monkey liver, but not that of the rat liver (Büchi *et al.*, 1974; Hsieh *et al.*, 1977*a*). Second, the ability to activate AFB_1 is found to be greater in the enzyme preparation from the rat liver than from the human liver. Third, the opposite is true with respect to the ability to detoxify AFB_1 (Hsieh *et al.*, 1977*a, b*).

The worst case daily intake of AFM_1 in the USA of 4·4 ng/kg body weight/day could also be compared with the daily intake of AFB_1, that gives an increment in lifetime risk of human liver cancer. Carlborg (1979) calculated a risk of 10 death/100 000, due to exposure to AFB_1 at 1 ng/kg body weight/day, on the basis of epidemiological studies in South-east Asia and Africa. However, the validity of this risk value for the USA was seriously doubted by Stoloff (1983, 1986). He carried out a study with a regional population that very likely had been exposed to aflatoxin levels in the same range as the highest exposures in Africa or Asia (13–197 ng/kg body weight/day). This regional population had essentially the same lifetime risk of liver cancer as a similar population ingesting AFB_1 at 0·2–0·3 ng/kg body weight/day. He further pointed out a number of methodological flaws in the epidemiological studies and questioned the strong correlation between the incidence of liver cancer and AFB_1. If the correlation between liver cancer and AFB_1 is tenuous, the correlation between liver cancer and AFM_1 must be even weaker.

Hsieh *et al.* (1986*b*) pointed out that the risk assessment as presented was based on data obtained from either adult animals or adult humans. The significance of the risk attributable to ingestion of AFM_1 by infant human populations would need to be evaluated separately, because milk represents a major constituent of the infant diet and because infant animals have been found to be considerably more susceptible to AFB_1 carcinogenicity than adult animals (Vesselinovitch *et al.*, 1972).

4.4 Conclusion

The limited animal studies carried out to determine toxicity and carcinogenicity of AFM_1 tend to come to the same qualitative conclusion: AFM_1 has hepatotoxic and carcinogenic properties. Quantitatively considered, the toxicity of AFM_1 in ducklings and rats seems to be similar or slightly less than that of AFB_1. The carcinogenicity is probably one to two orders of magnitude less than that of the highly carcinogenic AFB_1, as shown in studies with rainbow trout and rats and in genotoxicity studies. Taking account of both the carcinogenic potency and the levels of exposure, the cancer risk in the USA posed by AFM_1 is estimated to be much less than the cancer risk posed by AFB_1. It must be emphasized here that a quantitative characterization of AFM_1 carcinogenicity is a precarious matter, in view of the limitations of the animal studies that have led to this conclusion. The fact that AFB_1 is among the most potent carcinogens known, however, warrants concern about AFM_1 in dairy products.

5 REGULATION

5.1 Introduction

The hazard that mycotoxins may present to humans or livestock has led many countries to establish measures to control the contamination of foodstuffs and animal feedstuffs with mycotoxins. An enquiry made in 1987 at the request of the Food and Agriculture Organization (FAO), showed that there were 56 countries with known mycotoxin legislation (Van Egmond, 1989). All these countries had at least aflatoxin regulations for foodstuffs and/or animal feedstuffs, some had regulations for other mycotoxins as well. The institution of regulations for aflatoxins in animal feedstuffs is not only directed at protecting the health of the animal or limiting economic damage due to the adverse effect that some mycotoxins may have on animal productivity. A major reason is the protection of the health of the consumer of dairy products that are contaminated with AFM_1. Because of the direct link between AFB_1 in animal feedstuffs and AFM_1 in dairy products, attention in this section will be focussed to both these groups of products. Readers who wish to get an insight as well in the many regulations that exist for mycotoxins in foodstuffs other than dairy products, and for other mycotoxins in foodstuffs and animal feedstuffs, are referred to the complete publication (Van Egmond, 1989).

5.2 Aflatoxin Regulation for Animal Feedstuffs

5.2.1 Factors influencing the establishment of regulatory criteria
In addition to the justifications mentioned in Section 5.1 to establish aflatoxin regulations for animal feedstuffs, several other factors influence the decisions taken by authorities to establish acceptable levels for aflatoxins in certain animal feedstuffs at certain levels

(Schuller *et al.*, 1983; Park & Pohland, 1986). These include data on the occurrence and distribution of aflatoxins, the availability of analytical methodology and the regulations in other countries with which one trades.

Data on the occurrence of aflatoxins are needed to determine which commodities must be considered for regulatory action. These data will also allow an estimate of the effects of enforcement of regulations on the availability of feed and animal products. Data on the distribution of aflatoxins over the product(s) are indispensable in the establishment of regulatory criteria. If such a distribution is non-uniform, as is the case with aflatoxins in several commodities, there is a good chance, that the aflatoxin concentration in the lot to be inspected is wrongly estimated, due to the difficulties in representative sampling. High demands on sampling procedures will be necessary then.

To make inspection of commodities possible, accurate methods of analysis have to be available. The simplicity and validity of the methods will influence the amount and the reliability of data that will be generated and the practicality of the ultimate control measures taken. The limit of detection of the method will influence regulation, because a tolerance cannot be lower than the actual limit of detection of the method of analysis used.

Finally, the regulations in force in countries with which one trades have to be considered. If possible, the legislation under consideration should be brought into harmony with these existing regulations. Unnecessarily strict regulative actions may create difficulties for importing countries in obtaining supplies of animal feedstuffs. For exporting countries difficulties may arise in finding markets for their products.

There is no simple formula for weighing these factors in the decision-making process. Actually not much is made public about the rationales behind the enforcement of acceptable levels for aflatoxins in feedstuffs, except for the USA (Park & Pohland, 1986). Despite the dilemmas, aflatoxin regulations for feedstuffs have been established in the past decades in many countries. Because of the scope of this book, which deals with dairy products, the limits and regulations on aflatoxins in feedstuffs and feedstuff ingredients other than for dairy cattle will not be discussed.

5.2.2 Current limits and regulations
In Table 3 acceptable levels, legal bases, responsible authorities, status and references of methods of sampling and analysis that exist

TABLE 3
Aflatoxins in animal feedstuffs and feedstuff ingredients for dairy cattle: maximum acceptable levels and status of methods of sampling and analysis (after Van Egmond, 1989)

Country	Commodity	Limit ($\mu g/kg$) B_1	Limit ($\mu g/kg$) $B_1 + B_2 + G_1 + G_2$	Legal basis	Responsible authority	Method of sampling Status	Method of sampling Published	Method of analysis Status	Method of analysis Type[a]	Method of analysis Published[b]	Remarks
Austria	All feeds		50	Futtermittelverordnung 1981, Bundesgesetzblatt 48/1981–17.03. 1981							Situation 1981[c]
Brazil	Peanut meal (export)		50	Resolução no 79-XIII do CONCEX							Situation 1977[d]
Canada	Livestock feeds		20	Feed Regulations Canada Gazette Part II, *117*, no. 14, 2824, 27.07.1983	Agriculture Canada	Official	Agriculture Canada Sampling Procedures, 01-03-110, 28.09.1982	Not official	HPLC	Cohen & Lapointe (1981)	
(Republic of) China	Feedstuff ingredients, peanut cake, peanut cake with hull, peanut meal, soy bean pomace		1000		Ministry of Health, Council of Agriculture and Local Authorities	Official	Chinese National Standards Methods	Official			No more than 4% in final product
Dominican Republic	Maize and maize products		30								Situation 1981

(continued)

TABLE 3—contd.

Country	Commodity	Limit (µg/kg) B_1	Limit (µg/kg) $B_1 + B_2 + G_1 + G_2$	Legal basis	Responsible authority	Method of sampling Status	Method of sampling Published	Method of analysis Status	Method of analysis Type[a]	Method of analysis Published[b]	Remarks
European Community (Belgium, Denmark, France, Federal Republic of Germany, Greece, Ireland, Italy, Luxembourg, The Netherlands, Portugal, Spain, United Kingdom)	Complementary feedstuffs (for dairy cattle)	10		EC directive 86/229/3-06. 1986 Official Journal EC L 189/40. 1986	Various	Official	EC directive 76/371/ 01.03.1976. Official Journal EC L102/8. 1976	Official	TLC	EC directive 76/372/ 01.03.1976. Official Journal EC L102/9. 1976	Some EC countries have extra regulations for feedstuff ingredients (France, Portugal, United Kingdom). They are mentioned separately. All EC-tolerances refer to a commodity moisture content of 12%
	Following feedstuff ingredients: copra, palmnut, cottonseed, babassu, maize and derived products	200		EC directive 86/354/21.07 1986. Official Journal EC 212/27. 1986							Regulation coming into force 3.12.1988
France	Feedstuff ingredients	300		Journal Officiel de la République Française of 17.11.1981	Ministry of Agriculture; Ministry of Consumption						See also European Community
India	Peanut meal (export)	120		Licence under the solvent extraction oil, deoiled meal and edible flour (control) order 1967	Ministry of Food and Civil Supplies. Department of Civil Supplies	Official	Indian Standards Institution IS: 1714-1960 IS: 4115-1967				
Israel	Following European Community	Following European Community			Ministry of Public Health; Ministry of Agriculture			Official	TLC, qualitative filterpaper test	AOAC methods; Feedstuffs 23, 116. 1980	Situation 1981

Country	Commodity	Limit	Limit (2)	Legislation	Responsible authority	Legal status	Method	Reference	Remarks
Ivory Coast	Complete feedstuffs for dairy cattle	50			Ministry of Public Health, Ministry of Animal Production, Ministry of Commerce	Official			Proposal, types of aflatoxins not precisely stated
Japan	Peanut meal (import)	1000			Ministry of Agriculture Forestry and Fisheries	Official	Official TLC		Not more than 2% in feed for dairy cattle
Jordan	Animal feed	15			Ministry of Health				Situation 1981
Nigeria	Feedstuffs	50	30		Food and Drug Administration	Official	Official TLC	Stoloff & Scott (1984)	
Norway	Mixed feedstuffs	10–50 depending on type of animal			Ministry of Agriculture	Official			Groundnut meal and cottonseed meal are not allowed entry
Oman	Complete feedstuffs	10		Omanian Standard 46/1984 'Annex B: Prescribed Limit for harmful substances in compound Animal Feeds' Official Gazette 7/2/14054 of 01.11.1984	Ministry of Commerce and Industry	Official		Omanian Standard 48 'Methods of Sampling Animal Feeds'	Maximum content referred to a commodity moisture content of 12%
Peru	Mixed feedstuffs	20		Code of practice					
Poland	Feedstuff ingredients	50							
	Complete feedstuffs for dairy cows	20							
Portugal	Groundnut meal	500		Portaria no. 163/85. no. Diário da República de 23.03.1985	Ministry of Agriculture; Ministry of Public Health; Ministry of Commerce				See also European Community
	Other feedstuffs ingredients	50							

(continued)

TABLE 3—contd.

Country	Commodity	Limit (µg/kg) B_1	Limit (µg/kg) $B_1 + B_2 + G_1 + G_2$	Legal basis	Responsible authority	Method of sampling Status	Method of sampling Published	Method of analysis Status	Method of analysis Type[a]	Method of analysis Published[b]	Remarks
Romania	All feeds		<50	Joint papers vet. specialists and doctors of medicine. 1978. Veterinary sanitary journals. 1983	Ministry of Public Health; Ministry of Agriculture	Official	STAS protocol	Official	TLC	Bulletin of Information. 1978, 1979	
Sénégal	Peanut products (straight feedstuffs)	50		Journal Officiel de la République du Sénégal	Ministry of Commerce and Ministry of Public Health	Official		Official	TLC	EC-directive 76/372 01.03.1976. Official Journal EC L102/9, 1976	
	Peanut products (feeding-ingredients)	300									
Sweden	Raw materials	100		Landbruksstyrels forfattningssammling 1985: 35. Vb 10	Ministry of Public Health; Ministry of Agriculture				TLC	Stoloff & Scott (1984)	Tolerance for raw materials probably to be changed on 01.07.1987
	Complete feedstuffs for dairy cows										
Switzerland	Prohibit feeding cattle with peanut bruise			Futtermittelbuch and Schweizerisches Milchlieferungsregulativ 916.351.3 of 17.08.1977	Eidgenössische Forschungsanstalt für viehwirtschaft liche Produktion						
United Kingdom	Groundnuts or groundnut derivatives	50		The Feedingstuffs (no. 2) Regulations 1986. SI 1986 No. 1735	Ministry of Agriculture, Fisheries and Food; Local Authority Inspectors						See also European Community

					Official		Official	TLC, HPLC	Stoloff & Scott (1984)
United States	Feedstuffs and feedstuff ingredients	20	Food and Drug Administration compliance policy guide 7126.3 of 0815.82	Food and Drug Administration		Inspector's operating manual TN 82-11 and TN 83-18 of 05.05.1982 and 07.25.1983 respectively			
Uruguay	Cereals, oilseeds and feedstuffs	Dependent on type of animal	Decreto 529/980 sobre control de Calidad de alimentos para animales	Ministry of Agriculture					
Yugoslavia	Feedstuffs	Not given	Federal Committee for Agriculture						Situation 1981

[a]TLC = Thin Layer Chromatography; HPLC = High Performance Liquid Chromatography.
[b]If published outside the scientific literature, the full reference, where available, is given in the table.
[c]No information presented in 1986, data derived from enquiry of 1981.
[d]No information presented in 1986 and 1981, data derived from publication of Krogh (1977).

worldwide, are summarized for AFB_1 (and/or aflatoxins B_2, G_1 and G_2 where appropriate), in animal feedstuffs and feedstuff ingredients used for dairy cattle. The data presented reflect the situation as of 15 May 1987. They may not be complete or fully correct in a number of cases, due to problems experienced with language, terminology and inter-pretation of the responses to an international enquiry that the author undertook at the request of FAO (Van Egmond, 1989). Moreover some information originated from a previous enquiry of 1981 (Schuller et al., 1983). For any errors in his characterization of the current situation concerning aflatoxin regulations in feedstuffs for dairy cattle the author apologizes. The current limits for AFB_1 and for the sum of the aflatoxins B_1, B_2, G_1 and G_2 that specifically exist for (complete) feedstuffs for dairy cattle have also been represented in the form of frequency distributions in Figs 3 and 4 respectively.

In 1987, known actual and proposed legislation for aflatoxin(s) in animal feedstuffs existed in 34 countries. Of these, the countries of the European Community have a common directive. This directive had been tightened in 1984 when the acceptable level for AFB_1 in complementary feedstuffs for dairy cattle was reduced from 20 to 10 μg/kg. This reduction to 10 μg/kg concurs with the trend in Western European countries to establish limits for AFM_1 at a level of 0·05 μg/kg milk (see Section 5.3.2 and Fig. 5). Another recent development in

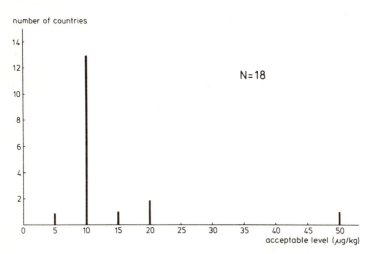

Fig. 3. Frequency distribution of acceptable levels (in force and proposed) of aflatoxin B_1 in feedstuffs for dairy cattle in various countries.

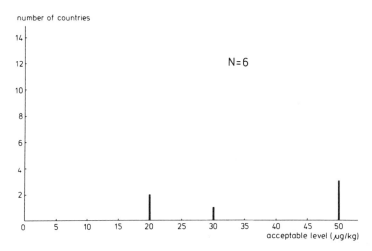

Fig. 4. Frequency distribution of acceptable levels (in force and proposed) of aflatoxins $B_1 + B_2 + G_1 + G_2$ in feedstuffs for dairy cattle in various countries.

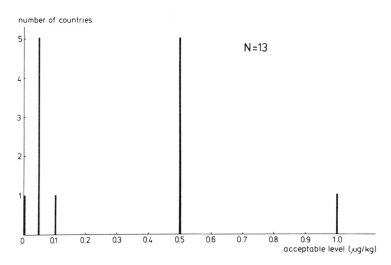

Fig. 5. Frequency distribution of acceptable levels (in force and proposed) of aflatoxin M_1 in milk in various countries.

Community Legislation is the introduction of an acceptable level for
AFB_1 in feedstuff ingredients at 200 $\mu g/kg$ by the end of 1988. Some EC
countries had already specified tolerances for some feedstuff ingredients,
as had several countries in other parts of the world. It is expected that
EC regulations for aflatoxins in animal feedstuffs will be further
tightened rather than relaxed in the near future.

   In addition to the common acceptable levels, the EC countries had
also adopted common directives for sampling and analysis. Only a few
countries outside the EC had indicated the use of official methods of
sampling and analysis. Most methods of analysis were based on Thin
Layer Chromatography (TLC). High Performance Liquid Chromatography (HPLC) methods occasionally were used, and immunoassays
were not mentioned at all in 1987. It may be expected however, that both
HPLC and Enzyme-Linked Immunosorbent Assay (ELISA) will
become of more importance, the latter especially for rapid screening of
feedstuff ingredients. Several commercial ELISA kits for aflatoxin
assay have become available already. However, at present ELISA
methods and systems still need intensive validation, before they can
become generalized tools of regulatory analysis.

5.3 Aflatoxin M₁ Regulation for Dairy Products

5.3.1 Factors influencing the establishment of regulatory criteria
The factors that influence the institution of regulations for AFM_1 in
dairy products are partly similar to those for feedstuffs.

   Countries that allow the use of feedstuff ingredients susceptible to
aflatoxin contamination must reckon with the possible contamination
of milk with AFM_1. The regulations in these countries for AFM_1 in milk
should bear relation with those for aflatoxins in feedstuffs for dairy
cattle.

   Data on the milk and dairy product consumption may influence the
decisions whether or not to establish regulations for AFM_1. Because the
distribution of AFM_1 in milk generally will be homogeneous, there is no
need for specially designed sampling plans. Although the distribution
within a product does not present problems, the partition of AFM_1
between dairy products is not even, as is comprehensively discussed in
Chapter 5. When specific criteria for milk products are established, the
fate of AFM_1 during processing must be taken into account, as well as
the specific problems that may arise from the analysis of the various
dairy products.

Special regulatory criteria may be desired for milk intended to be used for infant foods, because of the relatively high consumption rate of dairy products by infant foods and the possible high susceptibility of young individuals to aflatoxins (Vesselinovitch *et al.*, 1972). Finally the regulations in force in countries with which one trades may play a major role in the enforcement of regulations for AFM_1.

5.3.2 Current limits and regulations

In Table 4 acceptable levels, legal bases, responsible authorities, status and references of methods of analysis that exist worldwide, are summarized for AFM_1 in dairy products. The data reflect the situation as of 15 May 1987 and with respect to their correctness the same remarks are valid as for the feedstuffs data (see Section 5.2.2). The current acceptable levels for AFM_1 in liquid milk (not for specified milk-containing products, such as infant foods) are also presented in the form of a frequency distribution in Fig. 5.

In 1987, known actual and proposed legislation for AFM_1 in dairy products existed in 14 countries. Most of these countries had acceptable levels for specific products. For infant foods and children foods based on milk these limits were the lowest. The frequency distribution of Fig. 5 shows that for AFM_1 in milk two major tolerance peaks occurred, 0·05 and 0·5 μg/kg. It is indeed amazing to notice these large differences in acceptable levels between some Western European countries (0·05 μg/kg) and some American countries, the Soviet Union and Czechoslovakia (0·5 μg/kg). There is no firm scientific basis for establishing legal limits for AFM_1 (see Section 4) and it would be interesting to know the rationales behind the decisions to establish these limits. The international enquiry of 1986 did not include questionnaires about this.

A minority of the countries with AFM_1 regulations used official methods of analysis, in other cases the methods are still in the status of proposal. The analysis procedures were based on TLC and/or HPLC. Often AOAC methods were used, or methods derived from AOAC methods.

5.4 Reference Materials

The enforcement of regulations for aflatoxins in feedstuffs and AFM_1 in milk (products) requires monitoring of suspected products. Many laboratories perform large numbers of determinations of aflatoxins and consider themselves to be experienced and reliable. Nevertheless, Check Sample Programmes for aflatoxins in feedstuff ingredients as

TABLE 4

Aflatoxin M_1 in dairy products: maximum acceptable levels and status of methods of analysis (after Van Egmond, 1989)

Country	Commodity	Limit[a] (µg/kg)[a]	Legal basis	Responsible authority	Status	Type[b]	Published[c]	Remarks
Argentine	Fluid milk	$0.5\ (M_1 + B_1 + B_2 + G_1 + G_2)$		Ministry of Public Health, Ministry of Agriculture	Official		Stoloff & Scott (1984)	
	Milk-powder	$5\ (M_1 + B_1 + B_2 + G_1 + G_2)$			Official	TLC		
	Infant foods based on milk products	0.1			Proposal		Stubblefield (1979)	Proposal, to be considered by Comisión Nacional del Código Alimentario Argentino
Austria	Pasteurized fresh milk for infants and for children; children foods (calculated on reconstituted product)	0.01	Bundesgesetzblatt für die Republik Österreich 251. Verordnung of 14.05.86	Ministry of Public Health		TLC, HPLC	Arbeitsgruppe (1985)	
	Other milks and milk products (except the following):	0.05						
	Whey, liquid wheyproducts (except children foods)	0.025						
	Whey powder, whey paste (except children foods and calculated on dry matter)	0.4						
	Butter	0.02						
	Cheese	0.25						
	Powdered milk, milk powder products, condensed milk, milk concentrates (calculated on dry matter)	0.4						

Country	Product	(µg/kg)	Authority	Method	Reference	Notes
Belgium	Milk, powdered milk and condensed milk (reconstituted)	0·1	Ministry of Public Health		Koninklijk Besluit of 03.01.1975	Revision to 0·05 µg/kg expected
Brazil	Milk and milk products	0·5				Proposal
	Imported milk and milk products	0·1				Proposal
Czechoslovakia	All foods except the following:	5				
	Milk	0·5	Ministry of Health	Official	Ministerstro Zdravotnictvi ČSR Hygienické Předpisy 43/1978 supp. 61/1986	AOAC methods published in 'Acta Hygienica'
	Infant foods on milk basis (calculated on reconstituted product)	0·1				
	Other infant foods and children foods	1				
France	Milk-powder for infant foods	0·2		Official, TLC	Arrêté 20.10.1985 Journal Officiel de la République Française. 28.11.1985	
Federal Republic of Germany	Milk	0·05	Ministry of Public Health			Concept-proposal
	Milk for infant foods	0·01				
Nigeria	Fluid milk	1	Food and Drug Administration			

(continued)

TABLE 4—contd.

Country	Commodity	Limit[a] (µg/kg)[a]	Legal basis	Responsible authority	Status	Type[b]	Published[c]	Remarks
The Netherlands	Milk and milk products other than: Cheese	0·05		Ministry of Welfare, Public Health and Cultural Affairs				Proposal
	Butter	0·2						
		0·02						
	Milk-powder (calculated on milk-basis)	0·05	Landbouwkwaliteitsbesluit poedervormige melkprodukten Staatsblad no. 667 van het Koninkrijk der Nederlanden, 1982	Ministry of Agriculture, Ministry of Welfare, Public Health & Cultural Affairs	Proposal	TLC, HPLC	Stubblefield (1979) modified	
	Infant foods and milk for infant foods (calculated on milk-basis) (only for export)		Landbouwkwaliteitsregeling zuigelingenvoeding, Staatsblad no. 580 van het Koninkrijk der Nederlanden, 1984	Ministry of Agriculture, Ministry of Welfare, Public Health & Cultural Affairs				
Romania	Milk, dairy products	0	Joint papers veterinary specialists and doctors of medicine, 1978.	Ministry of Public Health, Ministry of Agriculture				
Sweden	Liquid milk products	0·05	National Food Administration's Ordinance Amending the Ordinance (SLV ES1985:1) on Foreign Substances in Food. SLV FS 1985:16	Swedish National Food Administration		HPLC	Stoloff & Scott (1984) De Vries & Chang (1982)	

Country	Product	Tolerance	Regulation	Authority	Status	Method[b]	Reference	Notes
Switzerland (existing regulations)	Milk, powdered milk and condensed milk (reconstituted) cream, butter milk	0·05	Verordnung über die hygienisch-mikrobiologische Anforderungen an Lebensmittel, Gebrauchs- und Verbrauchsgegenstände 817.024 of 14.09.1981	Laboratories of the Cantons	Official	TLC, HPLC	Arbeitsgruppe (1985)	Ordinance under revision. see hereafter
	Milk for infants or infant foods	0·01 ($M_1 + B_1$)						
	Whey, whey products (exclusive infant foods)	0·025						
	Butter	0·02						
	Cheese	0·25						
	Baby food	0·02						
(proposed regulations)	Milk and milk products	0·05		Laboratories of the Cantons				
	Cheese	0·25						
United States	Whole milk, low fat milk, skim milk	0·5	Food and Drug Administration Compliance Policy Guide 7106.10 of 10.01.1980	Food and Drug Administration	Official	TLC, HPLC	Stoloff & Scott (1984)	
Union of Socialist Soviet Republics	Milk and milk products	0·5	Methodic Documents Minszdrav USSR 2273-80 of 10.12.1980 and 4082-86 of 20.03.1986	Ministry of Health and State Agro-Industrial Committee of the USSR	Official		Methodic Documents Minszdrav USSR 2273-80 of 10.12.1980 and 4082-86 of 20.03.1986	
	Children foods	0						

[a] Specified if tolerance is relative to other (combinations of aflatoxins).
[b] TLC = Thin Layer Chromatography; HPLC = High Performance Liquid Chromatography.
[c] If published outside the scientific literature. the full reference. where available. was given.

peanut meal and maize meal and for AFM_1 in milk-powder have shown that large variability in results must be considered more as the norm than the exception (Friesen, 1982*a, b*). This fact gives little comfort to those who must either pay for the measurement or who base potentially important decisions upon them. It was this realization that prompted the Community Bureau of Reference (BCR) to undertake a Mycotoxin Programme in 1982 with the objective to improve the accuracy, and thereby the comparability of mycotoxin measurements. This is realised by the development of certified reference materials (CRM). Among the mycotoxins and matrices selected at the onset of the BCR Mycotoxin Programme in 1982 were milk-powders certified for their AFM_1 content, and peanut meal and compound feedstuffs certified for their AFB_1 content (Wagstaffe, 1987). The target concentrations of these CRM were selected, taking account of the current and proposed European legal limits (see Tables 3 and 4).

Priority was given to the development of the milk-powder reference materials. Four AFM_1 target concentrations were selected: zero, i.e. an effective analytical blank, $c.$ 0·1, $c.$ 0·3 and $c.$ 0·8 µg/kg milk-powder. These levels take into account the approximately eight-fold concentration that occurs during the spray drying of liquid milk. In 1986, three of the four milk-powder CRM have become available. Essential details of their development (preparation, stability and homogeneity, preliminary intercomparison and the steps taken to ensure accurate certification) have been published (Van Egmond & Wagstaffe, 1987). Certification of the fourth reference material (AFM_1 target concentration 0·1 µg/kg, a level chosen in view of the extreme demands in some European countries for milks for infant foods — see Table 4) was undertaken at a later stage, because of the additional difficulties presented in making accurate measurement at an AFM_1 content of 0·1 µg/kg. This material has become available at the end of 1987 (Van Egmond & Wagstaffe, 1988).

For AFB_1 in peanut meal, target concentrations of 0, 50 and 175 µg/kg were selected. For AFB_1 in compound feedstuffs target concentrations of 0 and 10 µg/kg were selected. At the time of writing, the development of the peanut meal and compounded feed CRM approached the stage of certification.

5.5 Conclusion

There are various countries with actual or proposed legislation on limits and regulations for aflatoxin(s) in feedstuffs for dairy cattle and AFM_1

in milk (products). The differences between tolerated levels sometimes vary widely between countries, particularly for AFM_1 in milk. Harmonization of these regulations would be highly desirable. Harmonization efforts should be supported by knowledge about the rationales behind the decisions that have led to the enforcement of the current regulations in the various countries of the world. The reliability of regulatory analysis data can be improved by systematically making use of certified reference materials. Milk-powder reference materials certified for their AFM_1 content have become available for this purpose through the European Community Bureau of Reference. Feedstuff (ingredient) reference materials, certified for their AFB_1 content are in the developing stage. The concentrations of aflatoxins in these reference materials take into account the current and proposed acceptable levels in the European Community.

6 SUMMARY

AFM_1 may occur in milk and milk products, resulting from the ingestion of AFB_1 in feedstuffs by the dairy cow. The excreted amounts of AFM_1 as a percentage of AFB_1 average 1–2%. This carry-over rate is approximately the same for high- and low-contaminated rations. It may vary from animal to animal, from day to day and from one milking to the next.

Surveillance programmes for AFM_1 in milk and milk products are carried out in many countries. In surveys carried out in the 1980s, AFM_1 levels were generally lower than those in the 1970s, and in many samples AFM_1 was not detectable.

In comparison with AFB_1 relatively little is known about the toxicity of AFM_1. The limited animal studies tend to come to the same conclusion that AFM_1 is a toxic and carcinogenic compound. Quantitatively considered, the toxicity of AFM_1 seems to be similar or slightly less than that of AFB_1. The carcinogenicity of AFM_1 is probably one or two orders of magnitude less than that of AFB_1. Taking account of both the carcinogenic potency and the levels of exposure, the cancer risk in the USA posed by AFM_1 is estimated to be much less than the cancer risk posed by AFB_1. The fact that AFB_1 is among the most potent carcinogens known, however, warrants concern about AFM_1 in dairy products.

Several countries control the contamination of dairy products with AFM_1 by regulation of the amounts of aflatoxins in feedstuffs for dairy cattle. Often, limits for AFB$_1$ at 10 µg/kg are applied. In addition, various countries have established regulations for AFM$_1$ on dairy products. The current range of (proposed) acceptable levels for AFM$_1$ in milk ranges from 0·05–0·5 µg/kg, with the exception of infant milk, for which lower levels exist.

The reliability of regulatory analysis data can be improved by systematically making use of certified reference materials. The European Community Bureau of Reference has available milk-powder reference materials, certified for their AFM$_1$ content at various levels. Dairy feedstuff reference materials, certified for their AFB$_1$ content are underway. The concentrations of aflatoxins in these reference materials take into account the current and proposed acceptable levels in the European Community.

ACKNOWLEDGEMENTS

The author wishes to thank the following persons for supplying data on the contamination of dairy products with AFM$_1$ in the 1980s: E. Brandl (Institut für Milchhygiene und Milchtechnologie, Vienna, Austria); H. Björk (Food Research Laboratory, Espoo, Finland); P. A. Burdaspal (Instituto Nacional de Sanidad, Majadahonda, Madrid, Spain); J. M. Frémy (Laboratoire Central d'Hygiène Alimentaire, Paris, France); J. Gilbert (Ministry of Agriculture, Fisheries and Food, Norwich, UK); W. Heeschen (Bundesanstalt für Milchforschung, Kiel, Federal Republic of Germany); E-L. Hintikka (The National Veterinary Institute, Helsinki, Finland); Liu Xingjie (Institute of Nutrition and Food Hygiene, Beijing, China); P. Mattson and T. Möller (National Food Administration, Uppsala, Sweden); S. Srebrnik (Institute of Hygiene and Epidemiology, Brussels, Belgium); A. Visconti (Instituto Tossine e Micotossine da Parassiti Vegetali, Bari, Italy); M. C. Walsh (The State Laboratory, Dublin, Ireland); Yan-Huei Wei (National Yang-Ming Medical College, Taipei, Republic of China).

The Food and Agriculture Organization of the United Nations is acknowledged for permission to reproduce parts of the information, collected in document ESN 803-10-1, prepared by the author of this chapter for the second FAO/WHO/UNEP Conference on Mycotoxins at Bangkok, Thailand, 28 September to 3 October 1987, and published at a later stage (Van Egmond, 1989).

Thanks are also due to J. R. Besling (Government Food Inspection

Service, Rotterdam, The Netherlands), D. P. H. Hsieh (University of California, Davis, California, USA), W. E. Paulsch and G. J. A. Speijers (National Institute of Public Health and Environmental Protection, Bilthoven, The Netherlands) for their helpful suggestions in preparing the manuscript.

Finally, thanks go to Ms Karen Kool for her careful preparation of the typescript for this chapter.

REFERENCES

Agricultural Advisory Committee for Environmental Critical Substances (1986). Annual report, Den Haag, The Netherlands (in Dutch).

Agricultural Advisory Committee for Environmental Critical Substances (1987). Annual report, Den Haag, The Netherlands (in Dutch).

Allcroft, R. & Carnaghan, R. B. A. (1963). Groundnut toxicity: an examination for toxin in human food products from animals fed toxic groundnut meal. *Vet. Rec.*, **75**, 259–63.

Allcroft, R. & Roberts, B. A. (1968). Toxic groundnut meal: The relationship between aflatoxin B₁ intake by cows and excretion of aflatoxin M₁ in milk. *Vet. Rec.*, **82**, 116–18.

Ames, B. N., McCann, J. & Yamasaki, E. (1975). Methods for detecting carcinogens and mutagens with the Salmonella mammalian microsome mutagenicity test. *Mutation Res.*, **31**, 347–64.

Applebaum, R. S., Brackett, R. E., Wiseman, D. W. & Marth, E. H. (1982). Aflatoxin: Toxicity to dairy cattle and occurrence in milk and milk products. A review. *J. Food Prot.*, **45**, 752–77.

Arbeitsgruppe Toxine 2 der Eidgenössischen Lebensmittelbuch-Kommission (1985). Amtliche Methoden zur Bestimmung von Aflatoxin M₁ in Milch und Milchpulver. *Mitt. Geb. Lebensm. Hyg.*, **76**, 92–103.

Björk, H. (1986). Pers. comm.

Boccia, A., Micco, C., Miraglia, M. & Scioli, M. (1985). A study on milk contamination by aflatoxin M₁ in a restricted area in Central Italy. Poster at VI Int. IUPAC Symp. on Mycotoxins and Phycotoxins, Pretoria, South Africa, 22–25 July, Abstr. P-65.

Borgström, S. (1983). Investigation of aflatoxin M₁ in milk and butter and organochloropesticides in milk. Swedish Dairy Association, Internal report SMRC-ö 107/a-2-1983 (in Swedish).

Brandl, E. (1986). Pers. comm.

Brown, C. A. (1982). Aflatoxin M in milk. *Food Technology in Australia*, **34**, 228–31.

Büchi, G. & Weinreb, S. M. (1971). Total synthesis of aflatoxins M₁ and G₁ and an improved synthesis of aflatoxin B₁. *J. Am. Chem. Soc.*, **93**, 746–52.

Büchi, G., Foulkes, D. M., Kurono, M., Mitchell, G. F. & Schneider, R. S. (1967). The total synthesis of racemic aflatoxin B₁. *J. Am. Chem. Soc.*, **89**, 6745.

Büchi, G., Muller, P. M., Roebuck, B. D. & Wogan, G. N. (1974). Aflatoxin Q₁: a major metabolite of aflatoxin B_1 produced by human liver. *Res. Commun. Chem. Pathol. Pharmacol.,* **8**, 585–91.

Büchi, G., Francisco, M. A., Liesch, J. M. & Schuda, P. F. (1981). A new synthesis of aflatoxin M_1. *J. Am. Chem. Soc.,* **103**, 3497–501.

Burdaspal, P. A., Legarda, T. M. & Pinilla, I. (1983). Note. Occurrence of aflatoxin M_1 contamination in milk. *Rev. Agroquin. Tecnol. Aliment.,* **23**, 287–90 (in Spanish).

Butler, W. H. & Barnes, J. M. (1968). Carcinogenic action of groundnut meal containing aflatoxin in rats. *Food Cosmet. Toxicol.,* **6**, 135–41.

Canton, J. H., Kroes, R., Van Logten, M. J., Van Schothorst, M., Stavenuiter, J. F. C. & Verhülsdonk, C. A. H. (1975). The carcinogenicity of aflatoxin M_1 in rainbow trout. *Food Cosmet. Toxicol.,* **13**, 441–3.

Carbonel, F. & Sudreau, P. (1981). Annual Report from Direction Générale de la Consommation, de la Concurrence et des Prix, Service de la Repression des Fraudes, Paris, France, pp. 66–8 (in French).

Carbonel, F. & Sudreau, P. (1984). Annual Report from Direction Générale de la Consommation, de la Concurrence et des Prix, Service de la Repression des Fraudes, Paris, France, p. 36 (in French).

Carlborg, F. W. (1979). Cancer, mathematical models and aflatoxin. *Food Cosmet. Toxicol.,* **17**, 159–66.

Christon, P., Miller, K. D. & Paan, A. (1985). Chemical conversion of aflatoxin B_1 to M_1. *Phytochemistry,* **24**, 933–5.

Cohen, H. & Lapointe, M. (1981). High pressure liquid chromatographic determination and fluorescence detection of aflatoxin in corn and dairy feeds. *J. Assoc. Off. Anal. Chem.,* **64**, 1372–6.

Cullen, J. M., Ruebner, B. H. & Hsieh, D. S. P. (1985). Comparative hepatocarcinogenicity of aflatoxins B_1 and M_1 in the rat. *Food Chem. Toxicol.,* **22**, 1027–8.

Cullen, J. M., Ruebner, B. H., Hsieh, L. S., Hyde, D. M. & Hsieh, D. S. P. (1987). Carcinogenicity of dietary aflatoxin M_1 in male Fisher rats compared to aflatoxin B_1. *Cancer Res.,* **47**, 1913–17.

De Iongh, H., Vles, R. O. & Van Pelt, J. G. (1964). Investigation of the milk of mammals fed on aflatoxin containing diet. *Nature (London),* **202**, 466–7.

De Vries, J. W. & Chang, H. L. (1982). Comparison of rapid high pressure liquid chromatographic and CB-methods for determination of aflatoxins in corn and peanuts. *J. Assoc. Off. Anal. Chem.,* **65**, 206–9.

Duncan, M., Brookes, P. & Dipple, A. (1969). Metabolism and binding to cellular macromolecules of a series of hydrocarbons by mouse embryo cells in culture. *Int. J. Cancer,* **4**, 813–9.

Ferrando, R., Palisse-Roussel, M. & Jacquot, L. (1984). Relay toxicity of powder milk aflatoxin. Medium term study on duckling. *CR. Acad. Sc. Paris* **298**, III-13, 355–8 (in French).

Frémy, J. M., Billon, J., Delpech, P. & Meurant, F. (1982*a*). Application of two bio-assays for aflatoxin M_1 toxicity evaluation in milk. *Rec. Méd. Vét.,* **158**, 461–6 (in French).

Frémy, J. M., Le Querrec, F. & Adroit, J. (1982*b*). In Annual report from Direction Générale de l'Alimentation, Service Vétérinaire d'Hygiène Alimentaire, Paris, France, p. 54 (in French).

Frémy, J. M., Le Querrec, F. & Adroit, J. (1983). In Annual report from Direction Générale de l'Alimentation, Service Véterinaire d'Hygiène Alimentaire, Paris, France, p. 58 (in French).

Frémy, J. M., Le Querrec, F. & Adroit, J. (1984). In Annual report from Direction Générale de l'Alimentation, Service Véterinaire d'Hygiène Alimentaire, Paris, France, p. 84 (in French).

Frémy, J. M., Le Querrec, F. & Adroit, J. (1985). In Annual report from Direction Générale de l'Alimentation, Service Véterinaire d'Hygiène Alimentaire, Paris, France, p. 42 (in French).

Friesen, M. D. & Garren, L. (1982a). International Mycotoxin Check Sample Program. Part I. Report on the performance of participating laboratories for the analysis of aflatoxins B_1, B_2, G_1 and G_2 in raw peanut meal, deoiled peanut meal and yellow corn meal. *J. Assoc. Off. Anal. Chem.*, **65**, 855–63.

Friesen, M. D. & Garren, L. (1982b). International Mycotoxin Check Sample Program. Part II. Report on the performance of participating laboratories for the analysis of aflatoxin M_1 in lyophilized milk. *J. Assoc. Off. Anal. Chem.*, **65**, 864–8.

Fritz, W., Donath, R. & Engst, R. (1977). Determination and frequence of aflatoxins M_1 and B_1 in milk and milkproducts. *Nahrung*, **21**, 79–84.

Gilbert, J., Shepherd, M. J., Wallwork, M. A. & Knowles, M. E. (1984). A survey of the occurrence of aflatoxin M_1 in UK-produced milk for the period 1981–1983. *Food Add. Contam.*, **1**, 23–8.

Green, C. E., Rice, D. W., Hsieh, D. P. H. & Byard, J. L. (1982). The comparative metabolism and toxic potency of aflatoxin B_1 and aflatoxin M_1 in primary cultures of adult-rat hepatocytes. *Food Chem. Toxic.*, **20**, 53–60.

Heeschen, W. (1986). Pers. comm.

Hintikka, E-L. (1986). Pers. comm.

Holzapfel, C. W., Steyn, P. S. & Purchase, I. F. H. (1966). Isolation and structure of aflatoxins M_1 and M_2. *Tetrahedron Letters*, **25**, 2799–803.

Hsieh, D. P. H. (1985). An assessment of liver cancer risk posed by aflatoxin M_1 in the Western world. In *Trichothecenes and other Mycotoxins*, ed. J. Lacey. John Wiley and Sons, New York, USA, pp. 521–8.

Hsieh, D. P. H., Wong, Z. A., Wong, J. J., Michas, C. & Ruebner, B. H. (1977a). Comparative metabolism of aflatoxin. In *Mycotoxins in Human and Animal Health*, ed. J. V. Rodricks, C. W. Hesseltine & M. A. Mehlman. Pathotox Publishers Inc. Park Forest South, IL, pp. 37–50.

Hsieh, D. P. H., Wong, J. J., Wong, Z. A., Michas, C. & Ruebner, B. H. (1977b). Hepatic transformation of aflatoxin and its carcinogenicity. In *Origins of Human Cancer*, ed. H. H. Hiat, J. D. Watson & J. A. Winsten. Cold Spring Harbor Laboratory, New York, pp. 697–707.

Hsieh, D. P. H., Beltman, L. M., Fukayama, M. Y., Price, D. W. & Wong, J. J. (1986a). Production and isolation of aflatoxin M_1 for toxicological studies. *J. Assoc. Off. Anal. Chem.*, **69**, 510–12.

Hsieh, D. P. H., Cullen, J. M., Hsieh, L. S., Shao, Y. & Ruebner, B. H. (1986b). Cancer risk posed by aflatoxin M_1. In *Diet, Cancer and Nutrition*, ed. K. Hayashi. Sci. Soc. Press, Tokyo, Japan; VNU, Utrecht, The Netherlands, pp. 56–65.

Hu Zhuohan, Xu Dadao, Hu Xiaohua, Wang Qing, Tang Beizhong, Weng

Congying, Yu Ming & Zhu Yuanzhen. (1985). Investigation of amount of aflatoxin M_1 consumed by infants of the urban area in Shanghai. *Acta Academiae Medicinae Primae Shanghai,* **12**, 38–40 (in Chinese).

Kiermeier, F. (1973a). Über die Aflatoxin-M-Ausscheidung in Kuhmilch in Abhängigkeit von der aufgenommenen Aflatoxin-B_1-Menge. *Milchwissensch,* **28**, 683–5.

Kiermeier, F. (1973b). Aflatoxin residues in fluid milk. *Pure Appl. Chem.,* **35** 271–3.

Kiermeier, F. & Mücke, W. (1974). Einfluss der Qualität des Futtermittels auf den Aflatoxin-gehalt der Milch. *XIX. Int. Milchw. Kongres ID,* 114–5.

Kiermeier, F., Reinhardt, V. & Behringer, G. (1975). Zum Vorkommen von Aflatoxinen in Rohmilch. *Dtsch. Lebensm.-Rdsch.,* **71**, 35–8.

Kiermeier, F., Weiss, G., Behringer, G., Miller, M. & Ranfft, K. (1977). Presence and content of aflatoxin M_1 in milk supplied to a dairy. *Z. Lebensm. Unters. Forsch.,* **163**, 71–4 (in German).

Kong Zhongfu, Dan Shuyin, Hu Wenjuan & Xin Xinjuan (1985). Studies on the contamination of food of animal origin with aflatoxin M_1. *Advances of Food Hygiene,* **3**, 68–73 (in Chinese).

Krogh, P. (1977). Mycotoxin tolerances in foodstuffs. *Pure Appl. Chem.,* **49**, 1719–21.

Lafont, P. & Lafont, J. (1975). Elimination d'aflatoxine par la mamelle chez la vache. *Cah. Nutr. Diét.,* **10**, 55–7.

Lafont, P. & Lafont, J. (1987). Génotoxicité des hydroxy-aflatoxines du lait. Abstract of lecture at Journées d'Etudes: Moisissures et Levures indesirable en Industrie Agro-Alimentaire, Paris, France, 25–27 November.

Lafont, P., Lafont, J., Mousset, S. & Frayssinet, C. (1980). Etude de la contamination du lait de vache lors de l'ingestion de faibles quantités d'aflatoxine. *Ann. Nutr. Alim.,* **34**, 699–708.

Lafont, P., Platzer, N., Siriwardana, M. G., Sarfati, J., Mercier, J. & Lafont, J. (1986a). Un nouvel hydroxy-dérivé de l'aflatoxine B_1: l'aflatoxine M_4. I. Production *in vitro* — structure. *Microbiol. Alim. Nut.,* **4**, 65–74.

Lafont, P., Siriwardana, M. G., Sarfati, J., Debeaupuis, J. P. & Lafont, J. (1986b). Un nouvel hydroxy-dérivé de l'aflatoxine M_4. II. Méthode de dosage — Mise en évidence de contaminations de laits commerciaux. *Microbiol. Alim. Nut.,* **4**, 141–5.

Lutz, W. K. (1978). *In vivo* covalent binding of aflatoxin B_1 and aflatoxin M_1 to liver DNA of pig and rat. In *Gesundheitsgefährdung durch Aflatoxine, Arbeitstag, Zürich,* ed. H. Poiger. Eigen Verlag Institut für Toxikologie der ETH und der Universität Zürich, Switzerland, CH-8603, pp. 65–71.

Lutz, W. K. (1979). *In vivo* covalent binding of organic chemicals to DNA as a quantitative indicator in the process of chemical carcinogenesis. *Mutation Res.,* **65**, 289–356.

Masri, M. S., Garcia, V. C. & Page, J. R. (1969). The aflatoxin M_1 content of milk from cows fed known amounts of aflatoxin. *Vet. Rec.,* **84**, 146–7.

McKinney, J. D., Cavanagh, G. C., Bell, J. T., Hoversland, A. S., Nelson, D. M., Pearson, J. & Selkirk, R. J. (1973). Effects of ammoniation on aflatoxins in rations fed lactating cows. *J. Am. Oil Chem. Soc.,* **50**, 79–84.

Mertens, D. R. (1979). Biological effects of mycotoxins upon rumen function and lactating dairy cows. In *Interactions of Mycotoxins in Animal Production*, Proceedings of a Symposium, National Academy of Sciences, Washington DC 20418, July 1978, pp. 118–36.

Ministry of Agriculture, Fisheries and Food. (1980). Survey of mycotoxins in the United Kingdom. The Fourth Report of the Steering Group on Food Surveillance, Working Party on Mycotoxins. HMSO, London, 14–17.

Möller, A. T. & Andersson, S. (1983). Aflatoxin M_1 in Swedish milk. *Vår Föda*, **35**, 461–5 (in Swedish).

Möller, A. T. (1987). Pers. comm.

Park, D. L. & Pohland, A. E. (1986). A rationale for the control of aflatoxin in animal feeds. In *Mycotoxins and Phycotoxins*, ed. P. S. Steyn & R. Vleggaar. Elsevier, Amsterdam, The Netherlands, pp. 473–82.

Patterson, D. S. P., Glancy, E. M. & Roberts, B. A. (1980). The carry-over of aflatoxin M_1 into the milk of cows fed rations containing a low concentration of aflatoxin B_1. *Food Cosmet. Toxicol.*, **18**, 35–7.

Piva, G., Pietri, A. & Carini, E. (1985). Occurrence of aflatoxin M_1 in dairy products. I. Italian milk and cheese. *Rivista della Società Italiana di Scienza dell 'Alimentazione*, **14**, 59–62 (in Italian).

Paul, R., Kalra, M. S. & Singh, A. (1976). Incidence of aflatoxins in milk and milk products. *Indian J. Dairy Sci.*, **29**, 318–22.

Polan, C. E., Hayes, J. R. & Campbell, T. C. (1974). Consumption and fate of aflatoxin B_1 by lactating cows. *J. Agric. Food Chem.*, **22**, 635–8.

Polzhofer, K. (1977). Aflatoxinbestimmung in Milch und Milchprodukten. *Z. Lebensm. Unters. Forsch.*, **163**, 175–7.

Pong, R. S. & Wogan, G. N. (1971). Toxicity and biochemical and fine structural effects of synthetic aflatoxins M_1 and B_1 in rat liver. *J. Natl. Cancer Inst.*, **47**, 585–90.

Purchase, I. F. H. (1967). Acute toxicity of aflatoxins M_1 and M_2 in day old ducklings. *Food Cosmet. Toxicol.*, **5**, 339–42.

Riberzani, A., Castelli, S., Del Vo, A. & Pedretti, C. (1983). Aflatoxin M_1 in some milk-based foods. *Industrie Alimentari*, **22**, 342–6 (in Italian).

Rice, D. W. & Hsieh, D. P. H. (1982). Aflatoxin M_1: in vitro preparation and comparative in vitro metabolism versus aflatoxin B_1 in the rat and mouse. *Res. Commun. Chem. Pathol. Pharmacol.*, **35**, 467–90.

Rodricks, J. V. & Stoloff, L. (1977). Aflatoxin residues from contaminated feed in edible tissues of foodproducing animals. In *Mycotoxins in Human and Animal Health*, ed. J. V. Rodricks, C. W. Hesseltine & M. A. Mehlman. Pathotox Publishers Inc., Park Forest South, Illinois, pp. 67–79.

Schuller, P. L., Verhülsdonk, C. A. H. & Paulsch, W. E. (1977). Aflatoxin M_1 in liquid and powdered milk (evaluation of methods and collaborative study). *Zesz. Probl. Postepow Nauk Roln.*, **189**, 255–61.

Schuller, P. L., Van Egmond, H. P. & Stoloff, L. (1983). Limits and regulations on mycotoxins. In *Proceedings of the International Symposium on Mycotoxins*, Cairo, Egypt, 6–8 September 1981, ed. K. Naguib, M. M. Naguib, D. L. Park & A. E. Pohland. Food and Drug Administration, Rockville, Maryland, USA and National Research Centre, Cairo, Egypt, pp. 111–29.

Sieber, R. & Blanc, B. (1978). Zur Ausscheidung von Aflatoxin M_1 in die Milch und dessen Vorkommen in Milch und Milchprodukten — eine

Literaturübersicht. *Mitt. Gebiete Lebensm. Hyg.,* **69**, 477–91.

Seiber, S. M., Correa, P., Dalgard, D. W. & Adamson, R. H. (1979). Induction of osteogenic sarcomas and tumors of the hepatobiliary system in non-human primates with aflatoxin B_1. *Cancer Res.,* **39**, 4545–54.

Sinnhuber, R. O., Lee, D. J., Wales, J. H. & Ayres, J. L. (1968). Dietary factors and hepatoma in rainbow trout. II. Carcinogenesis by cyclopropenoid fatty acids and the effect of gossypol and altered lipids on aflatoxin-induced liver cancer. *J. Natl Cancer Inst.,* **41**, 1293–301.

Sinnhuber, R. O., Lee, D. J., Wales, J. H., Landers, M. K. & Keyl, A. C. (1974). Hepatic carcinogenesis of aflatoxin M_1 in rainbow trout *(Salmo gairdneri).* *J. Natl Cancer Inst.,* **53**, 1285–8.

Sirica, A. E. & Pitot, H. C. (1979). Drug metabolism and effects of carcinogens in cultured hepatic cells. *Pharmac. Rev.,* **31**, 205–28.

Srebrnik, S. (1981). Annual Report Institute of Hygiene and Epidemiology, Ministry of Health, Brussels, Belgium, p. 78 (in Dutch and French).

Srebrnik, S. (1982). Annual Report Institute of Hygiene and Epidemiology, Ministry of Health, Brussels, Belgium, p. 75 (in Dutch and French).

Srebrnik, S. (1986). Pers. comm.

Stoloff, L. (1980). Aflatoxin M in perspective. *J. Food Prot.,* **43**, 226–30.

Stoloff, L. (1983). Aflatoxin as a cause of primary liver-cell cancer in the United States: a probability study. *Nutr. Cancer,* **5**, 165–86.

Stoloff, L. (1986). A rationale for the control of aflatoxins in human foods. In *Mycotoxins and Phycotoxins,* ed. P. S. Steyn & R. Vleggaar. Elsevier, Amsterdam, The Netherlands, pp. 457–71.

Stoloff, L. & Scott, P. M. (1984). Natural Poisons. In *Official Methods of Analysis of the Association of Official Analytical Chemists,* 14th edn, ed. S. Williams. AOAC, Arlington, Virginia, USA, pp. 477–500.

Stubblefield, R. D. (1979). The rapid determination of aflatoxin M_1 in dairy products. *J. Am. Oil Chem. Soc.,* **56**, 800–2.

Stubblefield, R. D., Shannon, G. M. & Shotwell, O. L. (1970). Aflatoxins M_1 and M_2: preparation and purification. *J. Am. Oil Chem. Soc.,* **47**, 389–90.

Van der Linde, J. A., Frens, A. M., de Iongh, M. & Vles, R. O. (1964). Inspection of milk from cows fed aflatoxin-containing groundnut meal. *Tijdschr. Diergeneesk.,* **89**, 1082–8 (in Dutch).

Van Egmond, H. P. (1989). Current situation on regulations for mycotoxins. Overview of tolerances and status of standard methods of sampling and analysis. *Food Add. Contam.* **6**, 139–88.

Van Egmond, H. P. & Stavenuiter, J. F. C. (1985). Developments in mycotoxin research. *Tijdschr. Diergeneesk.,* **110**, 1002–7 (in Dutch).

Van Egmond, H. P. & Wagstaffe, P. J. (1987). The development of milkpowder reference materials certified for their aflatoxin M_1 content (Part I). *J. Assoc. Off. Anal. Chem.,* **70**, 605–9.

Van Egmond, H. P. & Wagstaffe, P. J. (1988). The development of milkpowder reference materials certified for their aflatoxin M_1 content (Part II): Certification of milk powder RM 283. *J. Assoc. Off. Anal. Chem.,* **71**, 1180–2.

Van Pée, W., Van Brabant, J. & Joostens, J. (1977). La détection et le dosage de l'aflatoxine M_1 dans le lait et le lait en poudre. *Rev. Agric. (Bruxelles),* **30**, 403–15.

Vesselinovitch, S. D., Mihailovich, N., Wogan, C. N., Lombard, L. S. & Rao, K. V. N. (1972). Aflatoxin B₁, a hepatocarcinogen in the infant mouse. *Cancer Res.,* **32**, 2289–91.

Visconti, A., Bottalico, A. & Solfrizzo, M. (1985). Aflatoxin M₁ in milk, in Southern Italy. *Mycotoxin Research,* **1**, 71–5.

Wagstaffe, P. J. (1987). Reference materials for mycotoxin analysis. The work of the Community Bureau of Reference (BCR). In *Proceedings Symp. Aflatoxins and Edible Peanuts,* Leatherhead, 3 October 1986, pp. 24–32.

Walsh, M. C. (1986). Pers. comm.

Wiley, M. & Waiss, A. C. (1968). An improved separation of aflatoxins. *J. Am. Oil Chem. Soc.,* **45**, 870–1.

Williams, G. M. (1977). Detection of chemical carcinogens by unscheduled DNA synthesis in rat primary cell cultures. *Cancer Res.,* **37**, 1845–51.

Wogan, G. N. (1973). Aflatoxin carcinogenesis. In *Methods in Cancer Research, 7,* ed. H. Bush. Academic Press, New York, pp. 309–344.

Wogan, G. N. & Paglialunga, S. (1974). Carcinogenicity of synthetic aflatoxin M₁ in rats. *Food Cosmet. Toxicol.,* **12**, 381–4.

Wong, J. J. & Hsieh, D. P. H. (1976). Mutagenicity of aflatoxins related to their metabolism and carcinogenic potential. *Proc. Natl. Acad. Sci.,* **73**, 2241–4.

Yau-Huei Wei (1986). Unpublished data from Food and Drug Bureau of Executive Yuan of Republic of China.

Chapter 3

Chromatographic Methods of Analysis for Aflatoxin M₁

R. D. Stubblefield

Northern Regional Research Center, Peoria, Illinois, USA

&

H. P. van Egmond

National Institute of Public Health and Environmental Protection, Bilthoven, The Netherlands

1. Introduction.. 58
2. Sample Extraction.. 59
   2.1 Solvent extraction
   2.2 Solid phase extraction (SPE columns)
       2.3 Conclusion
3. Clean-up 63
   3.1 Thin-layer chromatography
   3.2 Solvent partition
   3.3 Column chromatography
   3.4 Conclusion
4. Ultimate Separation, Detection and Quantitation.................... 67
   4.1 Thin-layer chromatography
   4.2 High performance liquid chromatography
   4.3 Conclusion
5. Validation and Comparison of Chromatographic Methods......... 85
   5.1 Collaborative studies
   5.2 Other interlaboratory studies
   5.3 Conclusion
6. Summary.. 89

1 INTRODUCTION

Allcroft & Carnaghan (1963) found that milk from cows fed aflatoxin-contaminated diets contained a toxic substance which produced lesions characteristic of aflatoxin-poisoning in day-old ducklings. Soon, efforts were undertaken to detect and determine the 'milk-toxin', which was given the trivial name aflatoxin M. De Iongh *et al.* (1964) showed that aflatoxin M was a blue-violet fluorescent compound on thin-layer chromatography (TLC) plates under UV-radiation with a R_f well below that of aflatoxin B_1. The actual characterization of aflatoxin M was accomplished by Holzapfel *et al.* (1966) who showed aflatoxin M to consist of two closely related compounds, aflatoxins M_1 (AFM_1) and M_2 (AFM_2) (see Fig. 1), — the 4-hydroxy derivatives of aflatoxins B_1 and B_2. Purchase & Steyn (1967) proposed that the fluorescence of aflatoxin M_1 was approximately three times that of aflatoxin B_1, and therefore, quantitative measurement of aflatoxin M_1 could be achieved by comparison with aflatoxin B_1 standards on TLC plates. Although Stubblefield *et al.* (1972) later showed that the relative fluorescence intensities of aflatoxins B_1 and M_1 are nearly equal on TLC plates, this was the beginning of the quantitative measurement of aflatoxin M_1.

Reliable AFM methodology became available around 1967. Most methods were developed for analyzing milk powders (Purchase & Steyn, 1967; Roberts & Allcroft, 1968; Masri *et al.*, 1968, 1969; Stubblefield *et al.*, 1972), until Jacobson *et al.* (1971) published a method for liquid milk. All methods of this period were lengthy, laborious, and designed to measure the high AFM_1 concentrations from animal feeding studies. Modifications of the method of Jacobson *et al.* (1971), and of Purchase & Steyn (1967) by McKinney (1972) and Stubblefield *et al.* (1973) led to an evaluation of these methods and those of Roberts & Allcroft (1968),

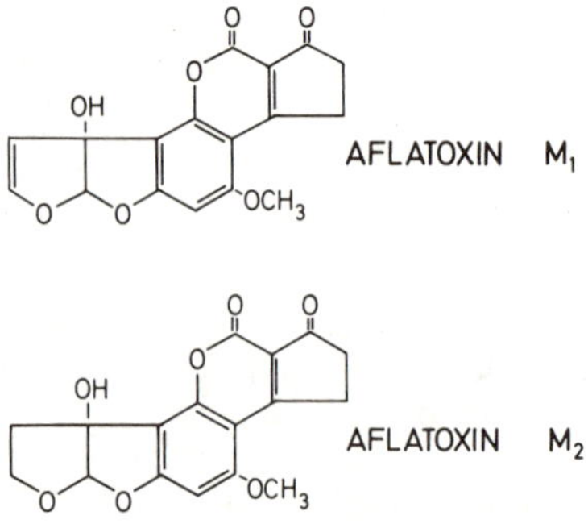

AFLATOXIN M_1

AFLATOXIN M_2

Fig. 1. Chemical structures of aflatoxins M_1 and M_2.

Masri *et al.* (1968, 1969), and Fehr *et al.* (1971). After an international collaborative study, the methods of Purchase & Steyn (1967), and Jacobson *et al.* (1971) were adopted as official methods by the Association of Official Analytical Chemists (AOAC) (1973) and the International Union of Pure and Applied Chemistry (IUPAC) (Purchase & Altenkirk, 1972; Purchase *et al.*, 1974). Pons *et al.* (1973) published a method that was refined by Stubblefield & Shannon (1974a) and which was applicable to a wide range of dairy products. This method became an official method in 1974 (Stubblefield & Shannon 1974b), and it permitted the analyst to use a single method for AFM$_1$ assays in powdered and liquid milks, creams, cheeses, and butter. This simplified assaying considerably, and it gave researchers a method that could detect AFM$_1$ at levels of 0·1 µg/kg milk.

The next major change came when Stubblefield (1979) published a rapid method for dairy products that eliminated the deproteinization step. This method is still an officially recognized method of IUPAC, AOAC, and the International Dairy Federation (IDF). The late 1970s also brought high performance liquid chromatography (HPLC) to aflatoxin assays. HPLC has become as common to AFM$_1$ measurements as TLC and densitometry, and more and more methods are being published with HPLC and fluorescence detection. The latter detection method has become a necessity because very low guidelines and tolerances in dairy products have been established for AFM$_1$ contamination (0·01–0·05 µg/kg in milk) in several Western European countries (see Chapter 2, Section 5.3). It is also imperative that samples be cleaned-up to a large degree if AFM$_1$ must be correctly identified at a level of 0.01 µg/kg (10 ppt) in a milk sample.

In this chapter, the authors will discuss the numerous problems and solutions associated with the chromatographic methods of analysis for AFM$_1$ in dairy products. These methods contain the basic steps that are part of each chemical analysis method for mycotoxins (see Fig. 2). The discussion of some aspects of extraction and clean-up is not restricted to chromatographic methods, but may also refer to immunochemical methods (see Chapter 4).

2 SAMPLE EXTRACTION

Food and feed products present their own particular problems for aflatoxin analyses. One problem of grains, oilseed crops, etc. which is not a concern with milk, is proper sampling. A uniform sample is easily

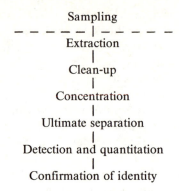

Fig. 2. Analytical procedure for mycotoxin determination.

obtained with milk because AFM_1 is distributed evenly throughout the fluid milk. No 'hot spots' occur in milk as in grains or other aflatoxin-contaminated solid substrates. Therefore, we do not have to be concerned with preparing a representative sample. The initial problem that is encountered in milk analyses is the extraction step. Milk is a complex natural product composed of sugars, proteins, minerals, vitamins, fats, etc., and consequently, AFM_1 is not easily extracted and purified for final assay. A process is needed that separates AFM_1 from milk easily, efficiently, and economically.

2.1 Solvent Extraction

Methods developed during the latter 1960s were primarily designed to analyze milk powders. Milk was spray dried or lyophilized to preserve the milk's shelf life and to reduce sample bulk. Various mixtures of methanol–water (Masri et al., 1968, 1969; Fehr et al., 1971), acetone-water and acetone–chloroform–water (Purchase & Steyn, 1967) were used to extract AFM_1 from the milk powder. Soxhlets (Purchase & Steyn, 1967; Fehr et al., 1971), blenders (Masri et al., 1968, 1969), and wrist-action shaker flasks (Purchase & Steyn, 1967) were used to extract the powder with the solvents. Jacobson et al. (1971) developed the first effective method for the determination of AFM_1 in fluid milks. Methanol–water was used as an extraction solvent. This method, modified by McKinney (1972) and Stubblefield & Shannon (1974a), became an official method of AOAC and IUPAC. All methods that use organic-aqueous mixtures to extract milk products incur the same problem: the need to include a deproteinization step. Some aqueous methanol mixtures will precipitate protein (casein) and simultaneously extract AFM_1, but in general, a precipitating agent must be used, e.g. lead acetate (Purchase & Steyn, 1967) and hydrochloric acid (Fehr *et al.*, 1971). The procedure requires time to accomplish and adds another step in which losses of AFM_1 can occur. Schuller *et al.* (1973) avoided the deproteinization step by mixing the milk with Celite® and packing the solid in a column. This approach is patterned after the early Celite® method developed for aflatoxin B_1 in peanuts (Nesheim *et al.*, 1964).

Stubblefield (1979) developed a method in which liquid milk was partitioned with chloroform in a separatory funnel (dichloromethane could also be used). Saturated sodium chloride was added to help prevent emulsions. This procedure eliminates the need to deproteinize the extract; however, an emulsion can occur with this extraction step, in particular with milk powders. The problem occurs occasionally and more often with some analysts than with others; therefore, the manner of funnel agitation must be important. Several proposed solutions have been put forth; heating, cooling, adding dodecylsulfate or salt, slow funnel rolling, etc., but none have been fully successful. Fukayama *et al.* (1980) extracted milk with a commercially available extraction column that contains an inert hydrophilic matrix (Chem Elut®, Analytichem International). This product is a granular diatomaceous earth, and the procedure is based on the same principles as the methods of Nesheim (1964) and Schuller *et al.* (1973). Milk is adsorbed on the column, and AFM_1 is eluted with chloroform–acetone. No emulsion will form with this procedure. It is necessary to completely evaporate the eluate to dryness because the presence of acetone will prevent AFM_1 from being adsorbed on a silica gel clean-up column. The method of Fukayama *et al.* (1980) was tested by the senior author (unpublished data), and the results of recovery studies were equal to those of the method of Stubblefield (1979).

The method of Fukayama *et al.* (1980) has been used for approximately six years, and it serves a definite purpose for those analysts who encounter emulsion problems with the method of Stubblefield (1979). Since the latter method has been recognized as an official method of the AOAC, IUPAC, and IDF, this alternative extraction procedure is important. Other scientists (Gauch *et al.*, 1979; Van Egmond *et al.*, 1982) have proposed similar methodology, which uses another commercially available extraction column (Extrelut®, Merck). The column is prepared with granular diatomaceous earth as is the one used by Fukayama (1980). In addition to milk, other aqueous matrices such as urine, blood, etc. can be adsorbed on these columns, and aflatoxins can be eluted as described.

2.2 Solid Phase Extraction (SPE Columns)

Reverse phase liquid chromatography uses C_{18}- or octyldecylsilane (ODS)-bonded silica gel columns to separate the aflatoxins for analytical or preparative purposes (Stubblefield & Shotwell, 1977; Beebe, 1978). Commercial companies capitalized upon this idea by manufacturing small columns (e.g. SEP-PAK®, etc.) containing approximately one gram of ODS material in disposable plastic cartridges. Winterlin *et al.* (1979) proposed the use of these columns to extract and concentrate AFM_1 from liquid milk, since the toxin is readily adsorbed to the bonded silica gel. No emulsion occurs, but the milk is usually diluted with water to pass the sample through the column cartridge faster. Another advantage of the C_{18} cartridges is that larger volumes of milk can be extracted by this procedure. This improves the minimum detection limit considerably, which is a major concern with AFM_1 analytical methodology. Several other researchers (Steiner & Battaglia, 1983; Ferguson-Foos & Warren, 1984; Takeda, 1984) have chosen to utilize the C_{18} cartridge extraction approach in their proposed methods, and it is becoming a very common, quick, and economical technique.

Although these methods vary slightly, they are generally simple. In a typical procedure, a C_{18} cartridge is prewashed with acetonitrile followed by water. The milk is passed into the cartridge, either *in vacuo* or by syringe. The cartridge is washed twice with water, and AFM_1 is eluted with a polar solvent, i.e. methanol, acetonitrile, acetone, or aqueous organic mixtures. The eluent is now ready for further clean-up. Some methods use the cartridge as a clean-up tool by passing other solvent mixtures through the cartridge prior to AFM_1 elution. Powdered milk can be redissolved in water (5 g milkpowder/50 ml H_2O), filtered, and treated as fluid milk. Cheeses can be extracted with methanol-water mixtures, diluted with water, and treated similarly; however, better recoveries are obtained with chloroform as an extraction solvent. Since extraction problems (emulsions) do not occur with cheeses, there is little need to change the recommended official methodology (Stubblefield, 1979).

2.3 Conclusion

There are three efficient procedures available to extract AFM_1 from fluid milk: (1) chloroform/dichloromethane partition (Stubblefield, 1979); (2) hydrophilic diatomaceous earth column partition followed by extraction with chloroform–acetone (Fukayama *et al.*, 1980); and (3) C_{18} cartridge adsorption followed by elution of AFM_1 with organic mixtures (Winterlin *et al.*, 1979). The first option is plagued occasionally with emulsion formation, but the other two options offer economical, quick alternatives.

3 CLEAN-UP

After the dairy sample has been extracted, the next important step is the partial purification or clean-up of the extract. This process is necessary to remove lipids and other substances that may interfere in the final detection or measurement by TLC or HPLC.

3.1 Thin-Layer Chromatography

Early techniques utilized the resolving power of TLC silica gel plates to purify extracts. Samples were developed on TLC plates, AFM_1 was eluted from the silica gel, and its concentration measured with a spectrometer. This technique is slow, laborious, and lacks sensitivity. It is a poor analytical procedure because elution from the silica gel is rarely quantitative. Also, AFM_1 is not obtained in sufficient quantities to detect low concentrations in samples.

3.2 Solvent Partition

Another technique to clean-up extracts prior to TLC or HPLC is solvent partition. Stubblefield *et al.* (1974*a*) washed an aqueous acetone extract of milk with hexane (or ether for anatto-colored cheese or butter), added NaCl, and extracted AFM_1 with chloroform for further clean-up on a cellulose column. A similar solvent partition procedure was followed by Tripet *et al.* (1981), but without the cellulose clean-up. Schuller *et al.*

(1973) used *n*-pentane to wash a methanol–water (+ NaCl) extract of milk, after which AFM$_1$ was re-extracted with chloroform for TLC. Gauch *et al.* (1979) used sodium hydroxide and hydrochloric acid to wash a dichloromethane/toluene extract of milk prior to TLC. Chambon *et al.* (1983) employed a washing step with hexane of a deproteinized milk filtrate, after which AFM$_1$ was extracted into chloroform. Chang & DeVries (1983) used petroleum ether to remove impurities from an acetonitrile extract of milk. A mixture of carbon tetrachloride and methanol–sodium chloride was used by Serralheiro & Quinta (1985) to clean-up powdered milk extracts. In the authors' experience, the solvent partition approach gives, in general, less clean extracts of milk than can be obtained with column clean-up procedures.

3.3 Column Chromatography

3.3.1 Laboratory-packed columns
Many methods commonly use silica gel column chromatography to clean-up the sample extracts for TLC or HPLC. Diatomaceous earth (packed) (Jacobson *et al.*, 1971) and cellulose (Pons *et al.*, 1973) columns have been used successfully in the past, but consistent, reliable data were not always achieved. Incidentally, other materials were used, such as alumina (Lafont *et al.*, 1981) and styrene divinylbenzene resin, a gel permeation matrix (Nijhuis *et al.*, 1983), but the use of these materials did not get wider application, and silica gel became the material used most often. Silica gel works well *if* the activity of the gel (water concentration) is carefully controlled. This fact became evident as methodology was developed for aflatoxin B$_1$ in agricultural commodities. If gel activities are not carefully controlled, laboratories may not get reproducible analytical results, and different product lots of silica gel may not give comparable data. Therefore, methodology-regulating organizations such as the AOAC, IUPAC and IDF specify the proper procedure to prepare silica gel activity for column chromatography. In practice, this is achieved by heating the silica gel to eliminate all absorbed water, after which a fixed amount of water is added to the silica gel, followed by equilibration for 24 h, and careful storage to keep the activity constant.

Once the silica gel is prepared, the next step is to prepare the column. The trend in mycotoxin methodology has been to use smaller columns and less silica gel. A small silica gel column is also employed in the official method of AOAC, IUPAC and IDF (Stubblefield, 1979). In this method, a glass column of 1·0 cm (i.d.) is used, equipped with a stopcock. Silica gel (2 g) is slurried in chloroform or dichloromethane and poured into the column. Excess solvent is drained through the stopcock to pack the column, and anhydrous sodium sulfate (granular) (1 g) is added to cap the column. A slight solvent excess is kept above the cap. It is important to know the preservative used in the chloroform or dichloromethane to prevent phosgene formation. Ethanol is the common preservative, and it is imperative the alcohol concentration does not exceed 1%. The authors have seen chloroform with levels of 2–3% ethanol, which will elute aflatoxins from silica gel. It is possible to obtain chloroform preserved with a nonpolar hydrocarbon. Use of this alternative avoids the problem.

The sample milk or dairy product extract should be dissolved in chloroform or dichloromethane and added to the packed column. The sample is drawn into the column by gravity. When the sample extract reaches the sodium sulfate cap, the column sides and sample extract container are rinsed with chloroform and drained into the column. The rinse step is repeated to ensure the entire sample is washed onto the silica column. The next step involves washing the column with a series of solvent mixtures to remove impurities from the sample extract. For the official method (Stubblefield, 1979) there are: acetic acid–toluene, hexane, and acetonitrile–ether–hexane. The washes are discarded, and AFM_1 is eluted with acetone–chloroform. The eluate is evaporated to prepare the final extract for measurement of AFM_1 by TLC or HPLC.

3.3.2 Silica gel and C_{18} cartridges

As mentioned earlier, smaller columns are desired in aflatoxin methodology. The commercially prepared so-called SPE columns (solid phase extraction) have become as popular as the columns prepared by scientists. These prepacked cartridges are convenient and economical. Their use as extraction columns was discussed earlier. Both silica gel cartridges (Frémy & Boursier, 1981; Cohen *et al.*, 1984; Tyczkowska *et al.*, 1984; Bijl & van Pethegem, 1985; Yousef & Marth, 1985; Bijl *et al.*, 1987) and C_{18} cartridges (Steiner & Battaglia, 1983; Cohen *et al.*, 1984; Ferguson-Foos & Warren, 1984; Hisada *et al.*, 1984; Takeda, 1984; Carisano & Torre, 1986) are used.

Two of these methods (Cohen *et al.*, 1984; Ferguson-Foos & Warren, 1984) make use of a combination of silica gel (cartridge and column respectively) and C_{18} cartridges. Of these, the method of Ferguson-Foos & Warren (1984) is simpler, and it produces a cleaner extract. In this procedure, milk is diluted with an equal volume of hot water (c. 80°C) and passed through a C_{18} SPE column. The column is washed with acetonitrile–water, and a vacuum is pulled on the cartridge to remove excess wash solution. After the cartridge is reprimed with 150 μl acetonitrile, AFM_1 is eluted with diethyl ether onto a silica gel column. An additional ether wash precedes elution of AFM_1 with dichloromethane–ethanol. The extract is evaporated, and the quantity of AFM_1 is measured by HPLC or TLC. This procedure is fast, economical, and efficient. It was subjected to an international collaborative study and accepted as an official AOAC method (Stubblefield & Kwolek, 1986). The clean-up accomplished by this method permits detection of AFM_1 at 50 ng/litre (ppt) milk easily, although the senior author has detected 10 ng/litre levels when an initial volume of 50 ml milk was used.

The advantages of the prepacked columns over the laboratory-packed columns are obvious. Variations in preparation of columns between analysts are eliminated, and the time needed to prepare the columns is saved. On the other hand, variations between lots of prepacked columns have been reported (Ferguson-Foos & Warren, 1984), and prepacked columns do not offer the possibility of easily introducing slight variations in the column composition (for instance, adjustment of the water content or column size).

3.3.3 Immunoaffinity cartridges

A new clean-up has become commercially available in the past 2–3 years. This column is composed of monoclonal antibodies, specific for AFM_1 (AFB_1 too), which are immobilized on Sepharose® and packed into small columns (see also Chapter 4, Section 3.3). These immuno-affinity columns have sufficient capacity (50–100 ng) to clean-up heavily contaminated samples. As of this writing, two companies produce the affinity columns (Cambridge/Noremco, Springfield, MO; Microtest/Oxoid, Basingstoke, England), but others will manufacture them, soon, no doubt. The immunoaffinity columns do an excellent job removing contaminants (Mortimer et al., 1987) because the AFM_1 antibody specifically recognizes AFM_1, so the column should not adsorb anything else. Consequently, very little, if any, impurities will be present in the final extracts for measurement by TLC, HPLC, or enzyme-linked immunosorbent assay (ELISA). This improves accuracy, precision, and sensitivity of assays. See Chapter 4, Section 3.3 for further details on immunoaffinity columns for AFM_1.

3.4 Conclusion

Several possibilities exist to separate AFM_1 from other co-extracted materials in crude extracts of milk and milk products. Solvent partition, is part of several procedures, but in general, it is less efficient than column chromatography. Both laboratory-packed silica gel columns and commercially available prepacked silica gel and C_{18} cartridges offer good possibilities in purifying extracts. The newest development is the immunoaffinity column, which offers the best potential for efficient dairy product clean-up.

4 ULTIMATE SEPARATION, DETECTION AND QUANTITATION

Extracts of dairy products that have been cleaned-up, are usually concentrated by evaporating the solvent in a rotary evaporator under reduced pressure, or by using a steam bath, while keeping the extract under a stream of nitrogen. The residue is redissolved in a small volume of solvent, suitable for chromatography, quantitatively transferred to a small vial, and brought to a specified volume. This concentration step is necessary to make detection and quantitation possible at the low concentration levels at which AFM_1 occurs in dairy products. Despite extraction and clean-up, the final extract may contain co-extracted substances possibly interfering with the AFM_1 determination. Several possibilities exist to separate AFM_1 from these matrix-components which permit detection and quantitation. Chromatographic procedures, which are physico-chemical separation techniques, are most often used, and are described hereafter. Immunoanalytical techniques, which are biochemical separation techniques, are described separately in Chapter 4.

Chromatographic processes involve solute partitioning between two phases: a stationary phase (the chromatographic bed) and a mobile phase (liquid or gas), that carries substances to be separated through the chromatographic bed. The stationary phase retards the progress of substances through the bed, depending on their physico-chemical properties, so that a separation into components can be achieved.

The chromatographic procedures that are used in the determinative steps in mycotoxin methods can be subdivided into:

— open-column chromatography
— thin-layer chromatography (TLC)

— high performance liquid chromatography (HPLC)
— gas–liquid chromatography (GLC)

Of these techniques, both TLC and HPLC are widely used in methods of analysis for AFM_1 and they are further discussed in the following sections.

Open-column chromatography has already been mentioned as a technique often used in clean-up procedures. A glass minicolumn with an internal diameter of $c.$ 5 mm, was used by Holaday (1981) to determine AFM_1 in milk. In that method, a minicolumn was prepared with successive zones of neutral alumina, sodium sulfate and Florisil®, with filter pulp packings at both ends. A toluene extract of milk is passed over the column, and AFM_1, if present, is trapped at the Florisil®–sodium sulfate interface as a blue-fluorescent band under UV-light. The limit of detection was reported to be 0·2 µg/litre milk, a value that could not compete with those obtainable with TLC and HPLC. The senior author consistently found a blue-fluorescent band when uncontaminated milk samples were tested. All attempts to remove the interfering contaminant were unsuccessful. Therefore, the minicolumn got no wider application.

GLC finds no application for AFM_1 determination, because AFM_1 is not volatile and would need to be derivatized before it can be gas chromatographed, whereas TLC and HPLC are readily applicable.

4.1 Thin-Layer Chromatography

In TLC, the stationary phase consists of a thin layer of adsorbent particles bound on a plate, e.g. glass, aluminium, or plastic mainly by spreading a suspension of sorbent to which a binder has been added. For mycotoxin assays, the thin layer usually consists of silica gel, because this type of adsorbent generally offers the best possibility of separating the toxin of interest from matrix components. Both precoated and laboratory-coated plates can be used. The characteristics of precoated and laboratory-coated plates may differ from brand to brand and even from batch to batch, leading to different separation behaviour.

Small volumes of sample extract are applied on the TLC plate by capillary pipettes or, preferably, microsyringes. After evaporation of the spotting solvent, the plate is placed in a separation chamber (developing tank) partially filled with the mobile phase (developing solvent). This developing solvent migrates through the thin layer by capillary forces and leads to separation into components, which can be made visible to make detection and determination possible. The displacement of the separated substances is expressed by their retardation factor (R_f), defined as the ratio of the migration of the compound and the migration of the solvent. In the first years of AFM_1 research, TLC became a common technique for separating AFM_1 from other extract components, after De Iongh *et al.* (1964) showed that AFM_1 could be detected as a blue-fluorescent compound on TLC plates under UV-light. Nowadays many official methods of analysis for AFM_1 determination still make use of this technique.

Initially there were no accurate standards of AFM_1 and there were no developing solvents known that could separate AFM_1 and AFM_2 on TLC. Therefore, Stubblefield *et al.* (1970) produced AFM_1 (and AFM_2) in analytical amounts through biosynthesis by a culture of *Aspergillus flavus* and purified the material for further physico-chemical characterization. Stubblefield *et al.* (1972) reported that on Adsorbosil®-1 TLC plates, an efficient TLC developing solvent for determination of AFM_1 is chloroform–acetone–isopropanol (85 + 10 + 5, v/v/v). Stubblefield *et al.* (1972) further found the ratio of fluorescence intensities of AFM_1 and AFB_1 on silica gel plates to be nearly 1, a value that differed significantly from the 3:1 ratio, earlier proposed by Purchase & Steyn (1967). Molar adsorptivities for AFM_1 were found to be 1995 m²/mol at λ_{max} (357 nm) in chloroform and 1882 m²/mol at λ_{max} (350 nm) in acetonitrile–benzene (10 + 90, v/v). The given molar absorptivity in chloroform has still been used by Van Egmond & Wagstaffe (1987, 1988) in studies on the preparation of milk powders with certified AFM_1 contents (see Chapter 2, Section 5.4). The AOAC recommends the molar absorptivity in acetonitrile–benzene for the preparation of analytical standard solutions. Purchase & Steyn (1972) investigated the storage behaviour of AFM_1 and concluded that the best way to store AFM_1 is to keep relatively large quantities in chloroform or acetonitrile–benzene solution in borosilicate glass.

In the early 1970s, several sources of reliable AFM_1 standards became available and various TLC methods for AFM_1 determination were published, which opened possibilities for many laboratories to undertake quantitative measurements of AFM_1 in milk and milk products. Mostly, TLC separations were carried out in one dimension with a single developing solvent (Purchase & Steyn, 1967; Roberts & Allcroft, 1968; Masri *et al.*, 1968, 1969; Fehr *et al.*, 1971; Jacobson *et al.*,

1971; McKinney, 1972; Stubblefield et al., 1972; Pons et al., 1973). Later, when pressures increased to lower the limits of determination, more analysts turned to two-dimensional TLC as a powerful separation tool. In the TLC determination of AFM_1, generally 10–20 μl of extract is applied to the plate. Depending on the desired accuracy and precision, different types of applicators are used. For screening purposes, disposable qualitative capillary pipettes will suffice. For quantitative work, disposable quantitative capillary pipettes or precision syringes, which are more accurate and precise, are used. Moreover, the latter allow the intermittent application of larger volumes under inert atmosphere by using them in combination with a repeating dispenser incorporated in a spotting device. The spotting of sample and standard is normally carried out according to a spotting pattern, prescribed as a part of the whole analytical procedure. Different spotting patterns apply to one-dimensional TLC and two-dimensional TLC.

4.1.1 One-dimensional TLC

In one-dimensional TLC procedures for AFM_1, aliquots of sample extract and AFM_1 standard solution are spotted next to each other (about 1·0–1·5 cm apart) on an imaginary line a few cm from the bottom of a 20 cm × 20 cm TLC plate. After development and subsequent drying, the TLC plate is observed under UV-radiation (λ: c. 365 nm) to make visual estimation and quantitation of AFM_1 spots possible. A useful property of AFM_1 (as the other aflatoxins) is that it emits the energy of absorbed (visible) longwave UV-radiation as visible blue-fluorescent light. This circumstance provides the basis for practically all the physical-chemical methods for AFM_1 detection and determination. AFM_1 quantities as low as 0·1 ng are visible on TLC plates with good quality UV-sources at a distance of 10 cm. High-intensity UV-sources make it possible to go to lower limits of detection. In Fig. 3, the results of one-dimensional TLC applied to extracts of liquid milk are shown. The extracts were prepared according to the method of Stubblefield (1979). The AFM_1 spots in the extracts can be located by comparison of R_f values and the fluorescence color with AFM_1 standard spots. Quantitation can be done by visual comparison of spot intensities, or instrumentally, with a densitometer. Visual comparison is subjective because the smallest detectable difference between aflatoxin spot intensities is about 20% (Beckwith & Stoloff, 1968), and the precision obtainable is about 15%. On the other hand, quantitation with a densitometer is objective, and can, at least in theory, give precision of better than ±5% (Nesheim, 1980).

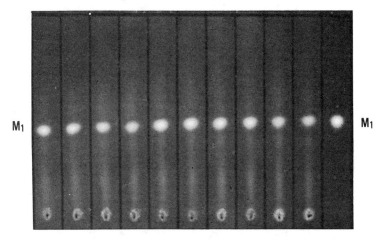

Fig. 3. One-dimensional TLC separation of milk extracts, prepared according to the method of Stubblefield (1979). Aflatoxin M₁ levels in samples are 0·5 μg/l milk. TLC conditions: 20 cm × 20 cm Macherey–Nagel silica gel plate, developing solvent: chloroform–acetone–isopropanol (87 + 10 + 3, v/v/v).

One-dimensional TLC is faster and easy to interpret. The lower limit of detection for AFM₁ in milk is around 0.1 μg/kg (0·1 ppb). If lower limits of detection are necessary (e.g. in those countries with acceptable limits for AFM₁ in milk of 0·05 μg/kg or less (see Chapter 2, Section 5.3.2), or when stubborn interferences in the extract are present that quench or enhance the fluorescence of the AFM₁ zone), two-dimensional TLC is the preferred separation technique. This avoids lengthy cumbersome clean-up steps.

4.1.2 Two-dimensional TLC

The application of two-dimensional TLC for AFM₁ determinations was introduced by Kiermeier & Mücke (1972) and Schuller *et al.* (1973). The technique has been incorporated in many other published methods for AFM₁ determinations in dairy products (Tuinstra & Bronsgeest, 1975; Patterson *et al.*, 1978; Stubblefield, 1979; Heeschen *et al.*, 1981; Nijhuis *et al.*, 1983; Gilbert *et al.*, 1984; Arbeitsgruppe Toxine 2, 1985; Bijl & Van Pethegem, 1985; Bijl *et al.*, 1987).

In Fig. 4 the spotting pattern for two-dimensional TLC as designed by Schuller *et al.* (1973) is shown. An aliquot of the sample extract is spotted at A and known amounts of AFM₁ standard are spotted at B. The plate is then developed in directions 1 and 2, respectively. The

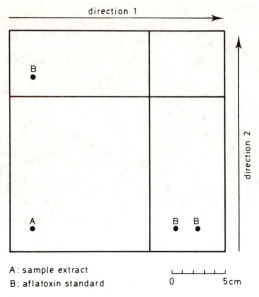

Fig. 4. Spotting pattern for two-dimensional thin-layer chromatography, after Schuller *et al.* (1973).

developing solvents in two-dimensional TLC must be compatible and independent, i.e. there should be little correlation between the retention patterns in both systems, otherwise the spots tend to agglomerate along the bisector of the plate. After complete development and drying, the TLC plate is inspected under longwave UV-radiation (365 nm). Figure 5 illustrates the two-dimensional separation pattern of an extract of milk powder, prepared according to the method of Stubblefield (1979), and modified as described by Van Egmond *et al.* (1986). With the help of the co-developed AFM_1 standard spots, the well-separated AFM_1 spot from the sample can be located. With a densitometer, the fluorescence intensities of the AFM_1 spots from sample and standard can be compared, and the AFM_1 concentration in the initial sample can be calculated.

If a densitometer is not available, an antidiagonal spotting pattern may be used as originally developed by Beljaars *et al.* (1973) for the determination of aflatoxin B_1 in peanuts (see Fig. 6). An aliquot of sample extract is spotted at A and different amounts of AFM_1 are spotted at the points B. The plate is developed two-dimensionally as just

Fig. 5. Two-dimensional TLC separation of milk powder extract, prepared according to the method of Stubblefield (1979). Aflatoxin M_1 level was c. 0·4 µg/kg milk powder. TLC conditions: 10 cm × 10 cm Merck silica gel plate; developing solvent 1: diethylether–methanol–water (94 + 4·5 + 1·5, v/v/v); developing solvent 2: chloroform–acetone–methanol (90 + 10 + 2, v/v/v).

described and detection and quantitation are again carried out under UV-radiation. With the help of the row of two-dimensionally developed AFM_1 standards, the AFM_1-spot from the extract can be located, and its concentration estimated by comparing its fluorescent intensity with that of the different AFM_1 standards. Since the AFM_1 spots are in a line and rather close to each other, the estimation is easier than visual comparison with standard spots developed in the side lanes, as in the densitometric spotting pattern. The antidiagonal technique has been incorporated in the methods of Tuinstra & Bronsgeest (1975), Nijhuis *et al.* (1983), Bijl & Van Pethegem (1985) and in the official Dutch procedures (Van Egmond *et al.*, 1986) for determination of AFM_1 in milk by visual estimation.

Two-dimensional TLC requires more time and materials than one-dimensional TLC, since only one sample at a time can be assayed by two-dimensional TLC. However, instead of the conventional 20 cm × 20 cm plates, 10 cm × 10 cm plates can be used (cut from the 20 cm × 20 cm plates). This size conserves developing time and materials and still gives excellent separations. The smaller plate offers the possibility

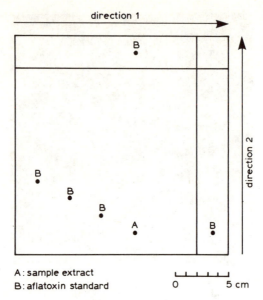

A: sample extract
B: aflatoxin standard

Fig. 6. Antidiagonal spotting pattern for two-dimensional thin-layer chromatography (visual quantitation).

of developing the plates in beakers. Several plates can be developed
simultaneously in the same beaker, which significantly reduces the
overall time required for a series of TLC runs.

4.1.3 Other TLC procedures

In addition to two-dimensional TLC, there exist other multiple
development techniques designed for the detection of low levels of
AFM_1 in extracts. They are all incorporated in Swiss procedures for the
determination of AFM_1, in milk. Three of these procedures (Gauch *et al.*
(1979), Steiner & Battaglia (1983) and the Method of the Cantonal
Laboratory Thurgau) have been published together (Arbeitsgruppe
Toxine 2, 1985) and have been given official status both in Switzerland
and in Austria. A fourth method has been published by Tripet *et al.*
(1981). In the method of Gauch *et al.* (1979), milk extracts and AFM_1
standards are applied adjacent to each other as for one-dimensional
TLC. Then the TLC plate is developed in the same direction 2–3 times
with various developing solvents. In the other procedures, milk extracts
and AFM_1 standards are applied next to each other near the middle of a

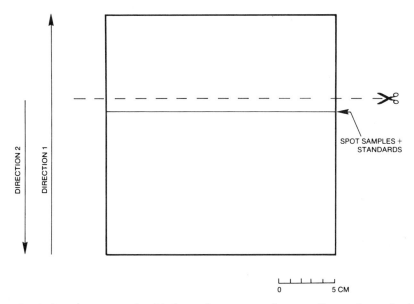

Fig. 7. Spotting pattern for thin layer chromatography according to the method of Cantonal Laboratory Zürich (Arbeitsgruppe Toxine 2, 1985). (Figure from Battaglia, 1988.)

TLC plate. (See Fig. 7.) Development is carried out in two directions at 180° to each other with three developing solvents (bottom-top, top-bottom, bottom-top). (In the method of Tripet *et al.* (1981), the third development is at 90° to the others.) This permits application of samples and standards with an automatic spotting device such as Camag Linomat®, and a higher precision with densitometry at the end of the chromatography.

The described multiple development procedures lead to well isolated zones of AFM_1 and make it possible to detect AFM_1 at levels down to $0·005 \, \mu g/kg$ milk.

4.1.4 Confirmation of identity

In spite of all the clean-up techniques used, there still may be substances in the final extract that behave in the same manner during TLC separation as AFM_1, even after multiple development TLC. In order to minimize the risk of false-positives, the identity of AFM_1 in positive samples needs to be confirmed. The most reliable method for this purpose is high resolution mass spectrometry (HRMS). HRMS in combination with TLC, however, is rather time-consuming, and not every laboratory is equipped with HRMS apparatus. Therefore, simpler techniques have to be available.

Probably the simplest way of confirming the presence of AFM_1 is the use of additional solvent systems, or the TLC procedure is repeated with the addition of AFM_1, as an internal standard, superimposed on the extract spot prior to developing the plate. After completion of TLC, this superimposed standard and the 'presumed' AFM_1 spot from the sample must coincide. Another possibility is to spray the developed TLC plate with a reagent; so that the color (of fluorescence) of AFM_1 changes. An example of the latter is the spraying test with a dilute solution of sulfuric acid which leads to a change in fluorescence of aflatoxin spots from blue to yellow (Smith & McKerman, 1962). Although the above described tests, if negative, would rule out the presence of AFM_1 they do not provide positive confirmatory evidence.

Positive identification can be obtained by formation of specific derivatives with altered chromatographic properties. Both AFM_1-standard and suspected sample are submitted to the same derivatization reaction. Consequently, in positive samples derivative(s) should appear, identical to derivative(s) from the standard.

Stack *et al.* (1972) applied this principle in a confirmatory test. This test was based on the formation of the acetate and the hemiacetal derivatives of presumptive AFM_1 in a clean extract (obtained by preparative TLC), and the TLC characteristics of the fluorescent reaction products were compared with the reaction products from AFM_1 standards. Because this procedure requires AFM_1 to be obtained through a preparative TLC step, Trucksess (1976) developed a simpler procedure, based on the in-situ reaction of trifluoroacetic acid (TFA) sample with AFM_1 directly on a TLC plate. In this procedure, small amounts of extract containing $1–2\,\mu g$ presumptive AFM_1 and AFM_1 standard solution are placed on a TLC plate and overspotted with TFA. After reaction at 55°C (which does not go to completion), the plate is developed and the R_f values of fluorescent spots of reacted and unreacted AFM_1 from the extract are compared with those of the AFM_1 standard, which has undergone the same procedure.

The method of Trucksess (1976) was further refined by Van Egmond *et al.* (1978), who modified the procedure such that derivatization with TFA is carried out on the TLC plate after two-dimensional development of the extract (see Fig. 5). TFA is spotted on the separated AFM_1 spot

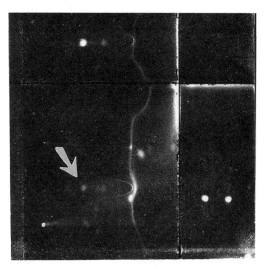

Fig. 8. Result of confirmatory test for aflatoxin M_1 after Van Egmond *et al.* (1978), applied to an extract of milk powder. Aflatoxin M_1 level was *c*. 0·4 μg/kg milk powder. TLC conditions: 10 cm × 10 cm Merck silica gel plate; two-dimensionally developed as described in Fig. 5. Developing solvent 3: chloroform–methanol–acetic acid–water (92 + 8 + 2 + 0·8, v/v/v/v).

from the extract and on the AFM_1 standard spots developed in the second direction. Reaction is carried out at 75°C, instead of 55°C, which shortens the reaction time. Then the plate is developed in the first developing direction and examined for blue-fluorescent reaction products. In Fig. 8, the result of such a confirmatory test, applied to an extract of milk, prepared according to the method of Stubblefield (1979) is shown. This modification of the method of Trucksess (1976) has the advantage that it can be carried out directly on the two-dimensionally developed TLC plate, that was used to quantitate AFM_1. AFM_1 quantities on the plate as low as 0·5 ng can be confirmed. The method of Van Egmond *et al.* (1978) was tested collaboratively by the AOAC and the IUPAC (Stubblefield *et al.*, 1980) and has been adopted as an official AOAC procedure (AOAC, *Official Methods of Analysis*, 1984). Although only 12 false negative results were reported from 212 samples, several collaborators encountered difficulties in spotting, zone diffusion, incomplete derivatization, or excessive time consumption.

To overcome these problems, the TFA spotting test was modified to a TFA–hexane spraying test (Van Egmond & Stubblefield, 1981). This

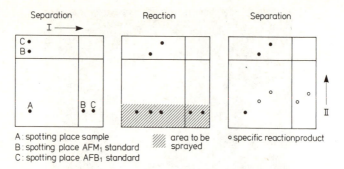

Fig. 9. Schematic representation of the two-dimensional confirmatory test for aflatoxins M_1 and B_1 after Van Egmond & Stubblefield (1981).

spraying test was based on the principles of the HCl spraying test of Verhülsdonk *et al.* (1977) for confirmation of identity of AFB_1 and AFG_1. It was designed such that it could also be used to confirm the identity of AFB_1 in animal tissue extracts. (See Fig. 9 for schematic presentation.) Extract and AFM_1 and/or AFB_1 standards are spotted on a TLC plate. (For milk extracts, AFB_1 standard can be left out of the spotting scheme, as it will normally not appear in milk.) The plate is developed two-dimensionally in mixtures of isopropanol–acetone–chloroform. After the first development, TFA–hexane is sprayed on that part of the plate containing the separated extract components and the undeveloped standard spots of AFM_1 (and AFB_1), and the plate is heated at 75°C. Then the plate is developed in a second direction, and the reaction products of AFM_1 (and AFB_1) with TFA from the extract are compared with the same derivatives of the respective standards. In Fig. 10 the result of this confirmatory test, applied to a milk extract containing AFM, and AFB, is shown. The method has been used successfully with extracts of milk, cheese, and liver containing AFM_1 (and AFB_1) at a level of 0·1 µg/kg and has been adopted as an official AOAC procedure (AOAC, *Official Methods of Analysis*, 1984).

4.2 High Performance Liquid Chromatography

In HPLC, the stationary phase consists of small adsorbent particles (typically 3–10 µm), densely packed in a tube (column). Columns are made of stainless steel, glass (fitted in a steel holder), or synthetic material.

In mycotoxin assays, the column packing consists of micro

Fig. 10. Results of confirmatory test for aflatoxin M_1 after Van Egmond & Stubblefield (1981), applied to an extract of milkpowder containing AFM, and AFB. Aflatoxin M_1 level was $0.2\,\mu g/kg$ milk. TLC conditions: 10 cm × 10 cm Merck silica gel plate: developing solvent 1: chloroform–acetone–isopropanol (82 + 10 + 8,v/v/v); developing solvent 2: chloroform–acetone–isopropanol (78 + 10 + 12, v/v/v).

particulate silica gel ('normal phase') or bonded phase packings, e.g. octadecylsilane (ODS; C_{18}) bonded to the silica support ('reverse phase'). Most columns that are used are prepacked, but there are satisfactory procedures available to self-pack the columns. Costs can be reduced by self-packing, but self-packed columns give less reproducible results than prepacked columns. The separation characteristics of column packings of the same type may differ from brand to brand.

Small volumes of sample extract (e.g. 10–100 μl) are introduced via the injector into the mobile phase (solvent) via a solvent delivery system, consisting of one or several high-pressure precision pumps. The solvent transfers the sample into the analytical column, which is usually protected by an easily interchangeable guard-column with the same type of packing material. Individual components are separated on the analytical column into chromatographic bands. The chromatographic bands are eluted from the column and passed through a detector. The detector signal is amplified and is traced on a recorder (integrating) in the form of chromatographic peaks that occur at certain retention times.

Mycotoxin peaks are identified by these retention times, which must correspond to those of injected standards. Quantitation can be done by comparison of peak area (or peak heights) of sample and standards.

HPLC became available for AFM_1 determinations in the late 1970s. Since then, HPLC has partly superseded TLC for AFM_1 determination, for various reasons. Separations are usually much better than those obtained with one-dimensional TLC; HPLC methods generally provide good quantitative information; and the equipment employed in HPLC systems can be automated rather easily. Today, many HPLC methods for AFM_1 have been published, although the procedural differences are often only slight (Glancy, 1979; Winterlin et al., 1979; Beebe & Takahashi, 1980; Frémy & Boursier, 1981; Gregory & Manley, 1981; Tuinstra & Haasnoot, 1982; Blanc et al., 1983; Chambon et al., 1983; Chang & De Vries, 1983; Cohen et al., 1984; Takeda, 1984; Tyczkowska et al., 1984; Yousef & Marth, 1985; Carisano & Torre, 1986). Of the published HPLC procedures for AFM_1, only the method of Ferguson-Foos & Warren (1984) makes use of normal phase systems, all other procedures are based on reverse phase systems. All HPLC methods for AFM_1 use fluorescence detection. UV-detectors are not suitable, because they are not selective and the obtained limits of detection for AFM_1 are relatively high, in the ng range.

4.2.1 Normal phase HPLC
In normal phase HPLC of AFM_1, columns are packed with silica gel, and chloroform-type solvents are used. Aliquots of sample extract and AFM_1 standard in chloroform (typically $100 \mu l$) are injected in the HPLC system, and after the separation on the column has occurred, AFM_1 is determined by fluorescence measurement (typical wavelengths for excitation and emission are at $\lambda = c.$ 360 nm and 430 nm respectively). Because AFM_1 does not exhibit strong fluorescence in normal phase solvents, a special provision in the flow-cell of the fluorescence detector is necessary to enhance the intensity of fluorescence of AFM_1. This is achieved by filling the cell with silica gel (Panalaks & Scott, 1977; Zimmerli, 1977).

In the adsorbed state, the aflatoxins B_1, B_2 and M_1 fluoresce much more intensively than they do in solution. The limits of detection thus obtained are of the same order of magnitude as those based upon TLC with fluorescence detection. The life expectancy of the packed flow-cell may vary depending on the number and state of the samples that are injected. In practice, contamination of the flow-cell occurs from

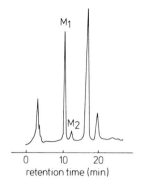

0   10   20
retention time (min)

Fig. 11. Silica gel normal phase HPLC separation of an extract of milk, prepared and chromatographed according to the method of Ferguson-Foos & Warren (1984). Aflatoxin M_1 and M_2 levels were 0·28 μg/kg and 0·075 μg/kg milk respectively. LC conditions: μ Porasil column, 0·4 cm × 25 cm (Waters); mobile phase: chloroform (22·5% water saturated)–ethanol (1000 + 22·5, v/v). (Figure from Ferguson-Foos & Warren, 1984.)

deposits and dirty extracts accumulating on the silica gel over a period of time. Therefore, it is necessary to change the silica from time to time. The latter restriction limits practical use of a packed flow-cell, especially when the detector has no easily accessible flow-cell. Panalaks & Scott (1977) indicated that it was not convenient to regenerate the flow-cell by pumping through a more polar solvent because this caused a change in the transparency of the silica gel.

In Fig. 11, a chromatogram of a normal phase HPLC separation of a milk extract is shown (Ferguson-Foos & Warren, 1984). Although AFM_1 levels as low as 0·03 μg/kg may be determined with this HPLC procedure, there are only a few laboratories that apply normal phase HPLC for AFM_1 determinations, because packing the flow-cell is inconvenient. Instead, reverse phase HPLC is preferred.

4.2.2 Reverse phase HPLC
Practically all columns used in reverse phase HPLC of AFM_1 are of the C_{18} type, except in the method of Chang & De Vries (1983), that uses a C_8 column. Solvents are usually combinations of aqueous buffers, acetonitrile and/or methanol (or isopropanol). The stationary phase is a nonpolar, bonded, silica gel. Consequently, the most polar compound is eluted first and the most nonpolar compound is eluted last ('reverse order'). Final extracts must be dissolved in water-miscible solvents.

Several of the reverse phase HPLC methods for AFM_1 determinations, use C_{18} SEP-PAK® cartridges for clean-up (see Section 3.3) resulting in aqueous final extracts that are readily injected into reverse phase systems. In procedures where the final extract is in chloroform, it is necessary to evaporate the extract and redissolve it in a water-miscible solvent (e.g. HPLC-solvent) before injection. Aqueous extracts of AFM_1 require particular attention because a study of AFM_1 methods revealed that AFM_1 is relatively unstable in water–acetonitrile (Van Egmond & Wagstaffe, 1987). It is, therefore, recommended that final extracts prepared in water–acetonitrile be injected shortly after their preparation and that AFM_1 standard solutions in this solvent mixture be freshly prepared.

Since AFM_1 fluoresces more intensely in reverse phase solvents than in normal phase solvents, packing the flow-cell with SiO_2 is not needed to obtain an acceptable fluorescent signal. Therefore, AFM_1 can be quantitated directly with fluorescence detection (typical wavelengths as for normal phase HPLC). In Fig. 12, a reverse phase chromatogram is shown of an extract of milkpowder, prepared according to the extraction and clean-up procedure of the method of Stubblefield (1979).

Although AFM_1 is readily detectable at low levels in reverse phase solvents, when employing sensitive fluorescence detectors, the fluorescence can be further increased 3–4 fold by derivatization with TFA (Beebe & Takahashi, 1980). (This reaction is also used in confirmatory tests for AFM_1, see Sections 4.1.4 and 4.2.3.) A disadvantage of derivatization is it requires an extra step in the analysis procedure. Beebe & Takahashi (1980) assumed the reaction product to be the hemiacetal of AFM_1, and it was tentatively designated AFM_{2a}; however, no evidence has been published that definitely identifies this derivative as the hemiacetal. These investigators were successful in the reproducible derivatization of AFM_1 to AFM_{2a}; however, other scientists have obtained incomplete derivative formation and have incorporated different reaction conditions (Gregory & Manley, 1981; Chang & De Vries, 1983; Hisada et al., 1984; Carisano & Torre, 1986).

Therefore, Stubblefield (1987) examined the reaction parameters for derivatization of AFM_1 with TFA. He concluded that the optimum conditions for forming AFM_{2a} from AFM_1 are mixing equal 200 μl portions of hexane and TFA with the dry dairy extract residue and allowing the mixture to react at 40°C for 10 min. It was also found that it is best to use sylilated vials for the derivatization of AFM_1 standards; however, sample extracts do not need sylilated vials. Reported limits of

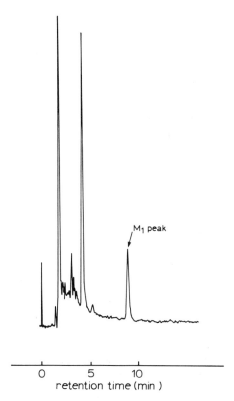

Fig. 12. C_{18} reverse phase HPLC separation of an extract of milkpowder, prepared according to the method of Stubblefield (1979), chromatographed as described by Van Egmond *et al.* (1986). Aflatoxin M_1 level was $0.4\,\mu g/kg$ milk powder. LC conditions: Polygosil 5-60 column (laboratory packed), (Macherey–Nagel): mobile phase: acetonitrile–water $(25 + 75, v/v)$. (Figure from Mulders, 1985.)

determination for AFM_1 in milk with reverse phase HPLC methods range from $0.005\,\mu g/kg$–$0.1\,\mu g/kg$. Although it is not clear from the publications how these limits of determination are defined, the ranges are generally comparable to those obtained with two-dimensional TLC.

4.2.3 Confirmation of identity

As is the case with TLC (see Section 4.1.4), coincidence of appearance and chromatographic properties in HPLC does not in itself provide proof that the isolated compound from the dairy product extract is chemically identical with the AFM_1 standard. Therefore, several of the HPLC methods for determining AFM_1 in dairy products have included a confirmatory step with TFA, when derivatization was not used to quantitate AFM_1 (see previous section).

In some of the methods, derivatization is carried out with the remaining extract and AFM_1 standard followed by TLC of the derivatives (Tuinstra & Haasnoot, 1982) or HPLC (Frémy & Boursier, 1981). In the procedure of Takeda (1984), the AFM_1 fraction from the extract is collected after HPLC and divided in two equal portions. One portion is re-injected without treatment with TFA, and the other is re-injected after treatment with TFA. Figure 13 shows the chromatograms of the non-treated AFM_1 fraction (A) and TFA-treated AFM_1 fraction (B) designated AFM_{2a}. Confirmatory proof of the presence of the AFM_1 in the extract fraction is indicated by the disappearance of the AFM_1 peak and the appearance of the AFM_{2a} peak in fraction B.

In HPLC procedures, where TFA-derivatization is used to quantitate AFM_1, confirmation of identity can be obtained by injecting underivatized extract and checking for absence of the AFM_{2a} peak in the chromatogram.

4.3 Conclusion

Both TLC and HPLC offer good possibilities to separate AFM_1 from matrix components in extracts of dairy products. In particular, multi-

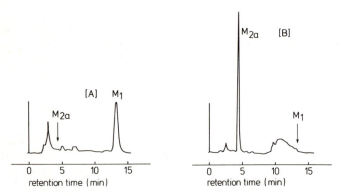

Fig. 13. C_{18} reverse phase HPLC separations of non-treated aflatoxin M_1 fraction (A) and TFA-treated aflatoxin M_1 fraction (B), designated AFM_{2a}, from milk extract, treated according to the method of Takeda (1984). Aflatoxin M_1 level was 1 μg/kg milk. LC conditions: μ Bondapak C_{18} cartridge, 8 mm $\times$ 10 cm (Waters): mobile phase: acetonitrile–methanol–water (15 + 15 + 70, v/v/v). (Figure from Takeda, 1984.)

development TLC, such as two-dimensional TLC, is a powerful separation tool, which offers the possibility of very low level determinations of AFM_1. Equipment for TLC is relatively cheap (except densitometers), and the technique can be learned relatively easily. On the other hand, HPLC procedures have become attractive because separations can be accomplished in a matter of minutes, and the equipment used in HPLC systems can be automated rather easily. In particular, reverse phase HPLC has become popular to determine AFM_1 in extracts of dairy products. The lower limits of detection for AFM_1 in dairy products, obtainable with two-dimensional TLC and HPLC, are comparable. Both techniques offer possibilities for confirmation of identity of AFM_1 in positive samples.

5 VALIDATION AND COMPARISON OF CHROMATOGRAPHIC METHODS

5.1 Collaborative Studies

The promulgation of regulations on AFM_1 in dairy products (see Chapter 2, Section 5.3) requires validated methods of analysis. Several international organizations (AOAC, IUPAC, IDF) spend time in collaborative testing of chromatographic methods of analysis for AFM_1. They validate the methods in typical applications and determine their practical and scientific characteristics. Among the practical characteristics are: cost of performance, time required, and level of training needed. Among the scientific characteristics are: accuracy, precision, specificity, and lower limit of detection. Although all these factors are of importance, the most important ones from the regulatory point of view are probably the accuracy and the between-laboratory variability (reproducibility).

The number of collaboratively studied methods of analysis for AFM_1, that have been published in the scientific literature, are limited. As yet, they are all based on TLC or HPLC. Most of these collaboratively studied methods were reviewed by Schuller *et al.* (1976), and by Scott (1989). It is inappropriate to repeat this reviewed information here, and the interested reader is referred to these reviews for further information.

Schuller *et al.* (1976) concluded that at AFM_1 levels in the order of $0.1-1$ μg/kg milk, the problem of separation of AFM_1 from the interferences resulted in between-laboratory coefficients of variation of 45–75%. After 1976, several collaborative studies were carried out for newer methods (both TLC and HPLC). The samples in these studies had AFM_1 levels ranging from approximately 0·03–1 µg/kg (expressed on the basis of fluid milk). These studies (Scott, 1989) yielded between-laboratory coefficients of variation of c. 40%, slightly lower than those of the methods reviewed by Schuller et al. (1976). These were obtained both for TLC (with densitometry) and HPLC. For regulatory arbitration, a coefficient of variation of about 40% has the following practical importance (Van Egmond et al., 1986): in countries where a tolerance of 0·05 µg/kg milk is proposed or in force (see Chapter 2, Section 5.3), the AFM_1 level in milk can be regarded as acceptable, as long as the result of a single analysis is lower than 0·1 µg/kg.

Collaborative tests of chromatographic methods of analysis for AFM_1 will continue to be undertaken in the near future because there is continuous development in AFM_1 analytical methodology. In particular, HPLC methods that include immunoaffinity clean-up will become candidates for collaborative study soon.

Despite the availability of collaboratively studied methods for AFM_1 determinations, many laboratories employ modifications of such methods or use other methods in their routine surveillance and quality control programmes. One reason is to increase the speed of analysis. The collaboratively studied methods are of particular importance because of their legal status.

5.2 Other Interlaboratory Studies

In addition to the collaborative studies, interlaboratory studies on AFM_1 methods have been conducted that permit a free choice of method. Published studies are the AFM_1 Check Sample Programme of the International Agency for Research on Cancer (IARC) in Lyon (Friesen & Garren, 1982), the Cooperative Study on the determination of AFM_1 at low levels in dried milk (Battaglia et al., 1984), and the intercomparison studies of methods of analysis for AFM_1 as carried out by the European Community Bureau of Reference (BCR) (Van Egmond & Wagstaffe, 1987, 1988). These studies served different purposes.

The Check Sample Programme of the IARC is an on-going Quality Assurance Programme whose primary objective is to provide laboratories, who perform AFM_1 determinations, a means of judging their analytical proficiency. In the published study (Friesen & Garren, 1982), lyophilized milk samples with an AFM_1 level of c. 8 µg/kg (corresponding to c. 1 µg/kg milk) were analyzed by 80 laboratories from 30 countries using a variety of methods. Participants were asked to give their best estimate of AFM_1 content in the samples. A between-laboratory coefficient of variation of 48% was obtained for all results, but it differed for each method group, ranging from 19–67%. The coefficient of variation of 19% occurred with a group of 12 laboratories using HPLC methods. This was a surprising result because these laboratories used a wide range of extraction and clean-up procedures. The authors stated that the quality of results grouped under a given method category could vary widely because the laboratories ranged from those just beginning aflatoxin assays to those who had routinely carried out such analyses for many years. It is possible that the laboratories who did HPLC analyses were more experienced than the others, which could have influenced the quality of the analysis results.

The cooperative study of Battaglia *et al.* (1984) was conducted to determine the comparability of analytical results obtained with various methods of analysis at the very low levels of AFM_1 contamination where statutory limits in several European countries have been issued (0·01–0·05 μg/kg milk, see also Chapter 2, Section 5.3). The study was carried out with milkpowders containing AFM_1 levels of 0, 0·1, and 0·5 μg/kg, corresponding more or less to AFM_1 levels of *c.* 0, 0·1, and 0·05 μg/kg liquid milk. It was conducted under auspices of the Working Party on Food Chemistry of the Federation of European Chemical Societies (FECS) and 58 participants from European countries took part. A total of 30 different methods were applied which utilized both TLC and HPLC. The results of the study showed that false positive results for the blank samples (31%) and false negative results for the contaminated samples (31%, sample 0·1 μg/kg; 3·4%, sample 0·5 μg/kg) were lower for experienced laboratories than for inexperienced laboratories. Similarly, the between laboratory coefficients of variation were lower for the experienced laboratories (54%, sample 0·1 μg/kg; 41%, sample 0·5 μg/kg) than for the inexperienced ones (72%, sample 0·1 μg/kg; 44%, sample 0·5 μg/kg). The occurrence of false-positives and false-negatives, and the variation coefficients seemed to be independent of the methods employed.

The BCR intercomparison studies consisted of two preliminary studies and two certification exercises. These were carried out in the framework of a project of the BCR on the preparation of milkpowder reference materials (RM) certified for their aflatoxin M_1 content (Van Egmond & Wagstaffe, 1987, 1988). (See also Chapter 2, Section 5.4.)

The preliminary studies (about 20 European participants) were conducted to identify and control the principal sources of analytical error, before a certification exercise was undertaken. Participants were asked to determine AFM_1 in a plain chloroform solution, in a chloroform extract of milkpowder, and in a milkpowder containing AFM_1. Although not in itself sufficient to allow certification, the design of the studies allowed the sources of error in the most divergent results to be identified and corrected before final certification: e.g. (1) inadequate resolution by some one-dimensional TLC systems of interferences and AFM_1; (2) problems arising from attempting to transfer the entire sample extract to the TLC plate; (3) decomposition of AFM_1 in acetonitrile–water used in the HPLC analysis. There were no significant method-dependent effects, although TLC–densitometry was noted to be at least as precise as HPLC, and TLC methods involving visual estimation did not allow sufficient precision for certification analyses.

The certification exercises were designed in the light of the preliminary studies. The first exercise (11 participants) led to the certification of three milkpowder RM; with AFM_1 contents <0·05 (RM 282), 0·31 (RM 284) and 0·76 (RM 285) respectively (Van Egmond & Wagstaffe, 1987). The second exercise (7 participants) led to the certification of milkpowder RM 283 (AFM_1 content 0·09 µg/kg). The 95% confidence intervals for RM 283, 284, and 285 were +0·04/−0·02, ±0·05, and ±0·06 µg/kg respectively. These were considered very good for the AFM_1 levels studied. The narrow divergence in the analysis results was attributed to the intensive preliminary studies, that had led to identification of important sources of error and that had given participants opportunities to increase their experience in accurately determining AFM_1 in milkpowder. Grouping of results according to characteristics did not indicate any systematic effect associated with the analytical method used.

5.3 Conclusion

Several methods for AFM_1 determination, based on both TLC and HPLC, have been collaboratively tested and performance characteristics of these methods have been established. They have particular value for arbitration of any possible dispute over AFM_1 levels in a specific commodity. Interlaboratory studies in which participants were given free choice of methods have shown, however, that experience in AFM_1 determination is at least as important as the application of a particular method or chromatographic technique to obtain accurate results.

6 SUMMARY

Many chromatographic methods of analysis have become available for the determination of AFM_1 in milk and milk products. Most of these were developed for the analysis of milk and milkpowder, but they can often be used for other dairy products as well, with minor modifications. Extraction of AFM_1 is usually accomplished by: (1) direct solvent extraction of the sample with chloroform or dichloromethane, or (2) indirect extraction from solid phase extraction columns, on which AFM_1 from the sample has been adsorbed, with more polar solvents as methanol and acetonitrile.

Extracts of dairy products are further purified by solvent partition, or more efficiently, by column chromatography. Both laboratory-packed silica gel columns and commercially available silica gel and C_{18} cartridges are successfully used. The newest development is the immunoaffinity column, which currently offers the best potential for efficient clean-up.

Purified extracts are concentrated to make detection of AFM_1 at low levels possible. Matrix components still present in final extracts can be separated from AFM_1 by thin-layer chromatography (TLC) or high performance liquid chromatography (HPLC). Several variants of these chromatographic procedures exist. One-dimensional TLC on silica-gel plates is the easiest application because it is a low-cost, fast separation technique. However, the technique is not suitable to determine AFM_1 at levels below $0.1\ \mu g/kg$ milk, and more advanced multidevelopment TLC techniques, such as two-dimensional TLC, are preferred in these cases. Another possibility to determine AFM_1 at very low levels is the application of HPLC, and in many laboratories, HPLC has superseded TLC. In particular, reverse phase HPLC has become popular in the determination of AFM_1. Lower limits of detection for AFM_1 in dairy products, obtainable with multidevelopment TLC and HPLC, are comparable.

Confirmation of identity of AFM_1 in positive samples can be achieved by derivatization of AFM_1 with TFA, followed by re-chromatography and inspection of the chromatograms for specific reaction products. This technique is applicable both for TLC and for HPLC.

Several methods for AFM_1 determination in dairy products have been validated through collaborative studies. They are of particular importance when legal action has to be taken. Interlaboratory studies, with free choice of method, have shown that experience in AFM_1

determination is at least as important as the choice of the method in obtaining accurate results.

ACKNOWLEDGMENTS

The authors thank J. R. Besling (Government Food Inspection Service, Rotterdam, The Netherlands) and W. E. Paulsch (National Institute of Public Health and Environmental Protection, Bilthoven, The Netherlands) for their helpful suggestions in preparing the manuscript.

The Typing Office of the National Institute of Public Health and Environmental Protection is gratefully acknowledged for the preparation of the typescript for this chapter.

REFERENCES

Allcroft, R. & Carnaghan, R. B. A. (1963). Groundnut toxicity: an examination for toxin in human food products from animals fed toxic groundnut meal. *Vet. Rec.,* **75**, 259–63.

Arbeitsgruppe Toxine 2 der Eidgenössischen Lebensmittelbuch-Kommission. (1985). Amtliche Methoden zur Bestimmung von Aflatoxin M_1 in Milch und Milchpulver. *Mitt. Gebiete Lebensm. Hyg.,* **76**, 92–103.

Association of Official Analytical Chemists (1973). Changes in Methods. *J. Assoc. Off. Anal. Chem.,* **56**, 485–6.

Association of Official Analytical Chemists (1984). *Official Methods of Analysis*, 14th Edn, ed. S. Williams. Association of Official Analytical Chemists, Arlington, VA, USA, 26.100, 26.107–26.109.

Battaglia, R., Van Egmond, H. P. & Schuller, P. L. (1984). Results of a FECS/ WPFC cooperative study on the determination of aflatoxin M_1 at low levels in dried milk. In *Proceedings Seminar Challenges to Contemporary Dairy Analytical Techniques*, Reading, UK, 28–30 March 1984, pp. 37–55.

Battaglia, R. Pers. comm. (1988).

Beckwith, A. C. & Stoloff, L. (1968). Fluorodensitometric measurement of aflatoxin thin layer chromatograms. *J. Assoc. Off. Anal. Chem.,* **51**, 602–8.

Beebe, R. M. (1978). Reverse phase high pressure liquid chromatographic determination of aflatoxins in foods. *J. Assoc. Off. Anal. Chem.,* **61**, 1347–52.

Beebe, R. M. & Takahashi, D. M. (1980). Determination of aflatoxin M_1 by high-pressure liquid chromatography using fluorescence detection. *J. Agric. Food Chem.,* **28**, 481–2.

Beljaars, P. R., Verhülsdonk, C. A. H., Paulsch, W. E. & Liem, D. H. (1973). Collaborative study of two-dimensional thin layer chromatographic analysis of aflatoxin B_1 in peanut butter extracts, using the antidiagonal spot application technique. *J. Assoc. Off. Anal. Chem.,* **56**, 1444–51.

Bijl, J. P. & Van Pethegem, C. H. (1985). Rapid extraction and sample clean-up for the fluorescence densitometric determination of aflatoxin M_1 in milk and milk powder. *Anal. Chim. Acta,* **170,** 149–52.

Bijl, J. P., Van Pethegem, C. H. & Dekeyser, D. A. (1987). Fluorimetric determination of aflatoxin M_1 in cheese. *J. Assoc. Off. Anal. Chem.,* **70,** 472–5.

Blanc, B., Lauber, E. & Sieber, R. (1983). Fixation de l'aflatoxine sur les proteines due lait. *Microbiol. Alim. Nutr.,* **1,** 163–77.

Carisano, A. & Torre, G. T. (1986). Sensitive reversed-phase high-performance liquid chromatographic determination of aflatoxin M_1 in dry milk. *J. Chromatogr.,* **335,** 340–4.

Chambon, P., Dano, S. D., Chambon, R. & Geachchan, A. (1983). Rapid determination of aflatoxin M_1 in milk and dairy products by high-performance liquid chromatography. *J. Chromatogr.,* **259,** 372–4.

Chang, H. L. & De Vries, J. W. (1983). Rapid high pressure liquid chromatographic determination of aflatoxin M_1 in milk and nonfat dry milk. *J. Assoc. Off. Anal. Chem.,* **66,** 913–7.

Cohen, H., LaPointe, M. & Frémy, J. M. (1984). Determination of aflatoxin M_1 in milk by liquid chromatography with fluorescence detection. *J. Assoc. Off. Anal. Chem.,* **67,** 49–51.

De Iongh, H., Vles, R. O. & Van Pelt, J. G. (1964). Milk of mammals fed an aflatoxin-containing diet. *Nature,* **202,** 466–67.

Fehr, P. M., Bernage, L. & Vassilopoulos, V. (1971). Effet la consommation de tourteau d'arachide pollue par *Aspergillus flavus* chez le ruminant en lactation. *Lait,* **48,** 377–91.

Ferguson-Foos, J. & Warren, J. D. (1984). Improved clean-up for liquid chromatographic analysis and fluorescence detection of aflatoxins M_1 and M_2 in fluid milk products. *J. Assoc. Off. Anal. Chem.,* **67** (1984), 1111–14.

Frémy, J. M. & Boursier, B. (1981). Rapid determination of aflatoxin M_1 in dairy products by reversed-phase high-performance liquid chromatography. *J. Chromatogr.,* **219,** 156–61.

Friesen, M. D. & Garren, L. (1982). International Mycotoxin Check Sample Programme. Part II. Report on the performance of participating laboratories for the analysis of aflatoxin M_1 in lyophilized milk. *J. Assoc. Off. Anal. Chem.,* **65,** 864–8.

Fukayama, M., Winterlin, W. & Hsieh, D. P. H. (1980). Rapid method for analysis of aflatoxin M_1 in dairy products. *J. Assoc. Off. Anal. Chem.,* **62,** 927–30.

Gauch, R., Leuenberger, U. & Baumgartner, E. (1979). Rapid and simple determination of aflatoxin M_1 in milk in the low parts per 10^{12} range. *J. Chromatogr.,* **178,** 543–9.

Gilbert, J., Shepherd, M. J., Wallwork, M. A. & Knowles, M. E. (1984). A survey of the occurrence of aflatoxin M_1 in UK-produced milk for the period 1981–1983. *Food Addit. Contam.,* **1,** 23–8.

Glancy, E. M. (1979). The analysis of milk for aflatoxin M_1 — a comparison of thin layer chromatography and high performance liquid chromatography methods. In *Proceedings 3rd meeting on mycotoxins in animal diseases,* ed. G. A. Pepin, D. S. P. Patterson & B. J. Shreeve. Weybridge, UK, April 1978.

Gregory III, J. F. & Manley, D. (1981). High performance liquid chromatographic determination of aflatoxins in animal tissues and products. *J. Assoc. Off. Anal. Chem.,* **64**, 144–51.

Heeschen, W., Blüthgen, A., Tolle, A. & Engel, G. (1981). Occurrence of aflatoxin M_1 in milk and dried milk in the Federal Republic of Germany. *Milchwissenschaft,* **36**, 1–4 (in German).

Hisada, K., Terada, H., Yamamoto, K., Tsubouchi, H. & Sakabe, Y. (1984). Reverse phase liquid chromatographic determination and confirmation of aflatoxin M_1 in cheese. *J. Assoc. Off. Anal. Chem.,* **67**, 601–6.

Holaday, C. E. (1981). Rapid screening method for aflatoxin M_1 in milk. *J. Assoc. Off. Anal. Chem.,* **64**, 1064–6.

Holzapfel, C. W., Steyn, P. S. & Purchase, I. F. H. (1966). Isolation and structure of aflatoxin M_1 and M_2. *Tetrahed. Lett.,* **25**, 2799–803.

Jacobson, W. C., Harmeyer, W. C. & Wiseman, H. G. (1971). Determination of aflatoxins B_1 and M_1 in milk. *J. Dairy Sci.,* **54**, 21–4.

Kiermeier, F. & Mücke, W. (1972). Uber den Nachweis von Aflatoxin M in Milch. *Zeitschr. Lebensm. Unters.-Forsch.,* **150**, 137–40.

Lafont, P., Siriwardana, M., Jacquet, J., Gaillardin, M. & Sarfati, J. (1981). Assay method for aflatoxin in milk. *J. Chromatogr.,* **219**, 162–6.

Masri, M. S., Page, J. R. & Garcia, V. C. (1968). Analysis for aflatoxin M in milk. *J. Assoc. Off. Anal. Chem.,* **51**, 594–600.

Masri, M. S., Page, J. R. & Garcia, V. C. (1969). Modification of method for aflatoxins in milk. *J. Assoc. Off. Anal. Chem.,* **52**, 641–3.

McKinney, J. D. (1972). Determination of aflatoxin M_1 in raw milk: a modified Jacobson, Harmeyer and Wiseman method. *J. Amer. Oil Chem. Soc.,* **49**, 444–5.

Mortimer, D. N., Gilbert, J. & Shepherd, M. J. (1987). Rapid and highly sensitive analysis of aflatoxin M_1 in liquid and powdered milks using an affinity column clean-up. *J. Chromatogr.,* **407**, 393–8.

Mulders, E. J. (1985). Pers. comm.

Nesheim, S. (1980). Factors affecting the thin layer chromatography of aflatoxins. In *Thin Layer Chromatography*, ed. J. C. Touchstone & D. Rogers. John Wiley and Sons, New York, pp. 194–240.

Nesheim, S., Barnes, D., Stoloff, L. & Campbell, A. D. (1964). Note on aflatoxin analysis in peanuts and peanut products. *J. Assoc. Off. Agric. Chem.,* **47**, 856.

Nijhuis, H., Heeschen, W. & Mühlhof, C. (1983). Experiments on the use of gel permeation chromatography in residue analysis. 2. Determination of aflatoxin M_1 in milk. *Milchwissenschaft,* **38**, 157–9 (in German).

Panalaks, T. & Scott, P. M. (1977). Sensitive silica gel-packed flowcell for fluorometric detection of aflatoxin by high-pressure liquid chromatography. *J. Assoc. Off. Anal. Chem.,* **60**, 583–9.

Patterson, D. S. P., Glancy, E. M. & Roberts, B. A. (1978). The estimation of aflatoxin M_1 in milk using a two-dimensional thin-layer chromatographic method suitable for survey work. *Fd Cosmet. Toxicol.,* **16**, 49–50.

Pons, Jr, W. A., Cucullu, A. F. & Lee, L. S. (1973). Method for the determination of aflatoxin M_1 in fluid milk and milk products. *J. Assoc. Off. Anal. Chem.,* **56**, 1431–6.

Purchase, I. F. H. & Alternkirk, B. A. (1972). Collaborative Study of the determination of aflatoxin M_1. *IUPAC Inform. Bull., Technical Report No. 6,* 9 pp.

Purchase, I. F. H. & Steyn, M. (1967). Estimation of aflatoxin M in milk. *J. Assoc. Off. Anal. Chem.,* **50**, 363–4.

Purchase, I. F. H. & Steyn, M. (1972). Fluorescence of aflatoxin M standards stored in solution and as dry films. *J. Assoc. Off. Anal. Chem.,* **55**, 1316–18.

Purchase, I. F. H., Stubblefield, R. D. & Altenkirk, B. A. (1974). Collaborative study of the determination of aflatoxin M₁ in milk. *IUPAC Inform. Bull., Technical Report No. 11,* 20 pp.

Roberts, B. A. & Allcroft, R. (1968). A note on the semi-quantitative estimation of aflatoxin M₁ in liquid milk by thin-layer chromatography. *Fd Cosmet. Toxicol.,* **6**, 339–40.

Schuller, P. L., Verhülsdonk, C. A. H. & Paulsch, W. E. (1973). Analysis of aflatoxin M₁ in liquid and powdered milk. *Pure Appl. Chem.,* **35**, 291–6.

Schuller, P. L., Horwitz, W. & Stoloff, L. (1976). A review of sampling plans and collaboratively studied methods of analysis for aflatoxins. *J. Assoc. Off. Anal. Chem.,* **59**, 1315–43.

Scott, P. M. (1989). Methods for determination of aflatoxin M₁ in milk and milkproducts — a review of performance characteristics. *Food Addit. Contam.* **6**, 283–305.

Serralheiro, M. L. & Quinta, M. L. (1985). Rapid thin layer chromatographic determination of aflatoxin M₁ in powdered milk. *J. Assoc. Off. Anal. Chem.,* **68**, 951–4.

Smith, R. H. & McKerman, W. (1962). Hepatotoxic action of chromato-graphically separated fractions of *Aspergillus flavus* extracts. *Nature,* **195**, 1301–3.

Stack, M. E., Pohland, A. E., Dantzman, J. G. & Nesheim, S. (1972). Derivative method for chemical confirmation of identity of aflatoxin M₁. *J. Assoc. Off. Anal. Chem.,* **55**, 313–14.

Steiner, W. & Battaglia, R. (1983). An efficient method for the determination of aflatoxin M₁ in milk and milkpowder in the lower ppt region. *Mitt. Gebiete Lebensm. Hyg.,* **74**, 140–6 (in German).

Stubblefield, R. D. (1979). The rapid determination of aflatoxin M₁ in dairy products. *J. Am. Oil Chem. Soc.,* **56**, 800–2.

Stubblefield, R. D. (1987). Optimum conditions for formation of aflatoxin M₁-trifluoroacetic acid derivative. *J. Assoc. Off. Anal. Chem.,* **70**, 1047–9.

Stubblefield, R. D. & Kwolek, W. F. (1986). Rapid liquid chromatographic determination of aflatoxins M₁ and M₂ in artificially contaminated fluid milks: collaborative study. *J. Assoc. Off. Anal. Chem.,* **69**, 880–5.

Stubblefield, R. D. & Shannon, G. M. (1974*a*). Aflatoxin M₁: analysis in dairy production and distribution in dairy foods made from artificially contaminated milk. *J. Assoc. Off. Anal. Chem.,* **57**, 847–51.

Stubblefield, R. D. & Shannon, G. M. (1974*b*). Collaborative study of methods for the determination of aflatoxin M₁ in dairy products. *J. Assoc. Off. Anal. Chem.,* **57**, 852–7.

Stubblefield, R. D. & Shotwell, O. L. (1977). Reverse phase analytical and preparative high pressure liquid chromatography of aflatoxins. *J. Assoc. Off. Anal. Chem.,* **60**, 784–90.

Stubblefield, R. D., Shannon, G. M. & Shotwell, O. L. (1970). Aflatoxins M₁ and M₂: preparation and purification. *J. Am. Oil Chem. Soc.,* **47**, 389–90.

Stubblefield, R. D., Shotwell, O. L. & Shannon, G. M. (1972). Aflatoxin M₁ and

M$_2$ and parasiticol: thin-layer chromatography and physical and chemical properties. *J. Assoc. Off. Anal. Chem.,* **55**, 762–7.

Stubblefield, R. D., Shannon, G. M. & Shotwell, O. L. (1973). Aflatoxin M$_1$ in milk: evaluation of methods. *J. Assoc. Off. Anal. Chem.,* **56**, 1106–10.

Stubblefield, R. D., Van Egmond, H. P., Paulsch, W. E. & Schuller, P. L. (1980). Determination and confirmation of identity of aflatoxin M$_1$ in dairy products: collaborative study. *J. Assoc. Off. Anal. Chem.,* **63**, 907–21.

Takeda, N. (1984). Determination of aflatoxin M$_1$ in milk by reversed-phase high-performance liquid chromatography. *J. Chromatogr.,* **288**, 484–8.

Tripet, F.-Y., Riva, C. & Vogel, J. (1981). Research of aflatoxins and determination of aflatoxin M$_1$ in milk products. *Trav. Chim. Aliment. Hyg.,* **72**, 367–79 (in French).

Trucksess, M. W. (1976). Derivatization procedure for identification of aflatoxin M$_1$ on a thin layer chromatogram. *J. Assoc. Off. Anal. Chem.,* **59**, 722–3.

Tuinstra, L. G. M. Th. & Bronsgeest, J. M. (1975). Determination of aflatoxin M$_1$ in milk at the parts per trillion level. *J. Chromatogr.,* **111**, 448–51.

Tuinstra, L. G. M. Th. & Haasnoot, W. (1982). Rapid HPLC method for the determination of aflatoxin M$_1$ in milk at the ng/kg level. *Fresenius Z. Anal. Chem.,* **312**, 622–3.

Tyczkowska, K., Hutchins, J. E. & Hagler, W. M. (1984). Liquid chromatographic determination of aflatoxin M$_1$ in milk. *J. Assoc. Off. Anal. Chem.,* **67**, 739–41.

Van Egmond, H. P. & Stubblefield, R. D. (1981). Improved method for confirmation of identity of aflatoxins B$_1$ and M$_1$ in dairy products and animal tissue extracts. *J. Assoc. Off. Anal. Chem.,* **64**, 152–5.

Van Egmond, H. P. & Wagstaffe, P. J. (1987). Development of milkpowder reference materials, certified for aflatoxin M$_1$ content (Part I). *J. Assoc. Off. Anal. Chem.,* **70**, 605–10.

Van Egmond, H. P. & Wagstaffe, P. J. (1988). Development of milkpowder reference materials, certified for aflatoxin M$_1$ content (Part II): Certification of milk powder RM 283. *J. Assoc. Off. Anal. Chem.,* **71**, 1180–2.

Van Egmond, H. P., Paulsch, W. E. & Schuller, P. L. (1978). Confirmatory test for aflatoxin M$_1$ on a thin layer plate. *J. Assoc. Off. Anal. Chem.,* **61**, 809–12.

Van Egmond, H. P., Sizoo, E. A., Paulsch, W. E. & Schuller, P. L. (1982). Rapid method of analysis for the semiquantitative determination of aflatoxin M$_1$ in milk. In *Handbook on rapid detection of mycotoxins.* OECD, Paris, pp. 38–47.

Van Egmond, H. P., Leussink, A. B. & Paulsch, W. E. (1986). The determination of aflatoxin M$_1$ in milk-powder, a comparison of four methods of analysis. *Bull. Int. Dairy Fed.,* **207**, 150–97.

Verhülsdonk, C. A. H., Schuller, P. L. & Paulsch, W. E. (1977). Confirmation reactions on aflatoxin B$_1$ and G$_1$. *Zeszyt. Probl. Postepow Nauk Rolniczyc,* **189**, 277–83.

Winterlin, W., Hall, G. & Hsieh, D. P. H. (1979). On-column chromatographic extraction of aflatoxin M$_1$ from milk and determination by reversed phase high performance liquid chromatography. *Anal. Chem.,* **51**, 1873–4.

Yousef, A. E. & Marth, E. H. (1985). Rapid reverse phase liquid chromatographic determination of aflatoxin M_1 in milk. *J. Assoc. Off. Anal. Chem.*, **68**, 462–5.

Zimmerli, B. (1977). Beitrag zur Bestimmung von Aflatoxinen mittels Hochdruck-Flüssigkeitschromatographie. *Mitt. Gebiete Lebensm. Hyg.*, **68**, 36–45.

Chapter 4

Immunochemical Methods of Analysis for Aflatoxin M₁

J. M. Frémy

Ministere de l'Agriculture, Laboratoire Central d'Hygiene Alimentaire, Paris, France

&

F. S. Chu

Department of Food Microbiology and Toxicology, Food Research Institute, University of Wisconsin, Madison, USA

1. Introduction 97
   1.1 Antigens and antibodies
   1.2 Types of immunochemical methods
   1.3 General protocol
2. Production and Characterization of Antibodies against Aflatoxin M₁ 101
   2.1 Preparation of immunogen
   2.2 Production and purification of polyclonal antibody
   2.3 Production of monoclonal antibody
   2.4 Antibody specificity
3. Immunochemical Procedures for Determination of Aflatoxin M₁ 109
   3.1 Radioimmunoassay (RIA)
   3.2 Enzyme immunoassay (EIA)
   3.3 Immunoaffinity chromatography (IAC)
4. Validation and Comparison of Immunochemical Procedures 120
   5. Summary 122

1 INTRODUCTION

1.1 Antigens and Antibodies

Immunoassays are based on the specific immunological reactions for the measurement for substances of biological or clinical interest (Voller & Bidwell, 1983). The principle of defense systems in higher vertebrates,

97

which aims at recognizing and then rejecting any invading foreign constituent (Daussant & Burreau, 1984), is used for the preparation of immunochemical reagents. As early as the late nineteenth century, Von Behring (Butler, 1980) found that 'humoral factors' which were capable of neutralizing, presumably through combining with toxins and micro-organisms, are present in the blood. By 1903, in recognizing the importance of these 'factors', a term of *'antibody/antibodies'* was introduced. The phagocytic cells' role in protecting the host body was also then demonstrated. Regardless of whether antibodies and/or phagocytic cells are the protecting factors, the substances that induce the defense action by producing such factors in the host are called *'antigens'*. A more detailed discussion on the chemical structure of antibodies will be treated later (Section 2.2.2).

Antigens are characterized by their *immunogenicity* as well as their *antigenicity*. Dependent on the size of the antigen to some extent, their immunogenicity resides in their capacity of inducing the formation of antibodies. Small molecules, such as aflatoxins, are not immunogenic and are called *'haptens'*. Once a hapten is conjugated to a large protein carrier, it becomes immunogenic. The antigenicity, or antigenic specificity, however, are characteristics/capabilities of the interaction of antigen with the antibodies. The interaction involves small sites, which are called *'antigenic determinants* or *epitopes'* on the surface of the antigen molecules (Daussant & Burreau, 1984).

1.2 Types of Immunochemical Methods

Although all the immuno-analytical techniques are based on the interaction between native antigens/haptens and specific antibodies, earlier applications were confined to the interaction between antibodies and large molecular weight antigens that lead to the precipitation of the antigen–antibody complex. The precipitation is used as a mean for the measurement of both the antigen or antibody concentration, and can occur both *in vivo* and *in vitro*. These procedures are suitable for measuring antigen concentrations in the μg–mg/ml range, and for the large molecular weight substances, such as proteins, only. Precipitates are not quantitatively formed when the concentrations for one of the reagents, either antigen or antibody, are in excess. Thus, other, more sensitive methods are necessary to measure the complex formation when the analyte is present at lower concentrations. In these cases, there is need not only for a good method for the separation of the free (unreacted) species from the complex (reacted), but also a sensitive marker antigen or antibody (labelled) for quantitation purpose. Generally, the terms *'immunochemical methods'* or *'immunoassays'* are used to refer to such modern techniques.

Most of the approaches used in the precipitation method for immunoassays are based on immunocomplex precipitation in agar gel which was originally developed by Oudin & Ouchterlony (Ouchterlony, 1949; Daussant & Burreau, 1984). Current precipitation methods for the analysis of food constituents, including such as the radial immunodiffusion (RID) of Mancini, the immuno-electrodiffusion (IED) of Laurell and the zone immunoassay (ZIA) of Vesterberg are summarized by Daussant & Burreau (1984).

Among different antibody or antigen/hapten labelling techniques used in immunoassay, the best known are fluoroimmunoassay (FIA), radioimmunoassay (RIA) and enzyme immunoassay (EIA) or enzyme-linked immunosorbent assay (ELISA). Other types of immunoassays which have undergone rapid development are: luminescent immunoassay (LIA) and metalloimmunoassay (MIA) (Voller & Bidwell, 1983). Immunoassays have been extensively used for large molecular weight proteins such as microbial toxins for many years (Oakley, 1970). Recent use of immunochemical methods for small molecular weight biologically active substances originated from the use of immunoassay for insulin (Yalow, 1978). These methods take advantage of two important phenomena: the potential specificity of antibodies to react with a given antigen, and the powerful amplification of the detectability of antigen-antibody complex achieved with labelling indicators (Kurstak, 1985).

Immunoadsorption methods are based on temporary immobilization of immunocomplexes on a solid support; immunoaffinity chromatography is a means of removing undesirable antibodies from an immune serum using the corresponding antigens. Conversely, immunoaffinity chromatography is also used for retaining a given antigen from a complex matrix by using the anti-antigen antibodies. Usually, the antigens or the antibodies are immobilized on a support and packed in a small column in the immunoaffinity chromatography.

Immunoassays for aflatoxins were not developed until the late 1970s for the following reasons:

(i)   Limitation due to the availability of antibody. Aflatoxins are haptens. Thus, it is necessary to conjugate them to a protein carrier before immunization for antibody production. Large amounts of toxins (several milligrams) are needed to prepare the conjugate. Such large quantities of aflatoxins were not commonly available.

(ii) Limitation due to the availability of sensitive markers. Labelling of aflatoxin (marker) is not very easy to do. Again, a large amount of toxin is necessary.

(iii) Necessity of high sensitivity. It is necessary to determine very low amounts of residues or metabolites of toxin in very complex matrices such as dairy products.

At present, only RIA, ELISA and immunoaffinity chromatography methods have been developed for aflatoxin determination.

1.3 General Protocol

Unlike many bacterial toxins which are highly immunogenic, aflatoxins are low molecular weight substances; hence, aflatoxins are not immunogenic and must be conjugated to a protein or polypeptide carrier before immunization. The development of immunoassay protocols is concentrated in four areas:

(i)   Preparation of immunogen. This includes the formation of a toxin derivative and conjugation of it to a protein for immunization.

(ii) Production of antibodies. This includes immunization, collection of immune serum and purification of antibodies.

(iii) Characterization of antibodies. This includes determination of antibody titer for a specific marker and antibody specificity. The cross-reactivity of the antibody with structurally-related aflatoxin compounds should be established.

(iv) Development of immunoassay (RIA, ELISA and immuno-affinity chromatography) protocols. This includes preparation of reagents (such as labelled indicators, affinity gel, etc.), development of assay conditions and protocols as well as practical application of such protocols for the analysis of commodities naturally contaminated with aflatoxins and collaborative studies (Chu, 1984*a*).

2 PRODUCTION AND CHARACTERIZATION OF ANTIBODIES AGAINST AFLATOXIN M_1

2.1 Preparation of Immunogen

2.1.1 Derivatization

Aflatoxins do not have a reactive group that can be used for direct coupling to proteins to elicit the immunogenic effect. Therefore, it is necessary to introduce a reactive group into the aflatoxin molecule by derivatization before conjugation. Several methods for the introduction of a reactive group in the aflatoxin molecule have been developed:

— A carboxymethyl oxime (CMO) group can be introduced at the carbonyl position in the cyclopentenone ring of aflatoxin M_1 (AFM_1) (Harder & Chu, 1979) as shown in Fig. 1.

Fig. 1. Preparation of immunogen for aflatoxin starting from either aflatoxin B_1 or M_1. Aflatoxin B_1 (1) or M_1 (2) was first converted to the oxime derivative (3) and then conjugated to the protein (4) (after Chu & Ueno, 1977, and Harder & Chu, 1979).

— Because large quantities of AFB_1 are more easily available than that of AFM_1, two other approaches were made using AFB_1. A carboxymethyl oxime–AFB_1 can be also prepared (Chu & Ueno, 1977) as shown in Fig. 1; or, AFB_1 can be converted to AFB_{2a} and then either directly conjugated to protein through the formation of a Schiff's base or converted to the hemiglutarate derivative (Lau *et al.*, 1981) as shown in Fig. 2 before conjugation.

— Another approach was developed by Sizaret & Malaveille (1983) who synthesized an AFB_1 8,9-dichloride (AFB_1–Cl_2) derivative as shown in Fig. 3.

Fig. 2. Preparation of immunogen for aflatoxin starting from aflatoxin B_{2a}. Aflatoxin B_1 (1) was first converted to aflatoxin B_{2a} (2) which was further converted to the hemiglutarate derivative (3) and then conjugated to the protein (4) (after Lau *et al.*, 1981).

Fig. 3. Preparation of immunogen for aflatoxin starting from aflatoxin B₁ dichloride. Aflatoxin B₁ (1) was first converted to the dichloride derivative (2) and then conjugated to the protein (3) (after Sizaret & Malaveille, 1983).

— Aflatoxin can also be conjugated to protein after conversion to aflatoxin Q₁ (Fan *et al.*, 1984).

2.1.2 Conjugation to protein

After derivatization to introduce a reactive group in the aflatoxin molecule, the toxin is then ready for conjugation to protein. In a review, Chu (1984*b*) summarized several extensive studies made by his laboratory and by others. Different aflatoxin derivatives can be conjugated to protein carriers using the water soluble carbodiimide or the mixed anhydride or the reductive alkylation methods (Chu, 1984*b*). Although bovine serum albumin (BSA) is the protein most frequently used as the carrier, other proteins such as ovalbumin (OVA), gamma globulin and keyhole limpet haemocyanin (KLH) have been used.

Since antibody specificity is determined primarily by the specific group exposed on the protein molecule, different aflatoxin analogs can be used for conjugation to prepare an immunogen which would elicit antibody of different specificity. For example, when AFM₁ was conjugated to the protein carrier through the cyclopentenone ring of the molecule, the —OH difuran group was exposed on the protein carrier;

thus, the antibody produced was most specific to AFM_1. However, when AFB_{2a} or AFB_1 8,9-dichloride was used as derivative for conjugation to the protein carrier, the cyclopentenone ring (common to AFB_1, AFB_2, AFM_1) is exposed on the protein carrier and the antibody produced bound almost equally with AFB_1 and AFM_1.

2.2 Production and Purification of Polyclonal Antibody

2.2.1 Immunization

Although immunization procedures vary slightly from laboratory to laboratory, the following factors need to be considered:

(i)   The sites and routes of immunization. This can be intravenous, intramuscular, subcutaneous or intraperitoneal in the back or/ and in thigh by multiple injections or at only a few sites.

(ii) The use of adjuvants for injection. The incomplete adjuvant acts as a depot of immunogen in the organism and the antigen/ hapten is emulsified with mineral oil before injection. The complete adjuvant enhances the humoral response and the antigen/hapten are emulsified with mineral oil and killed (inactivated) *Mycobacteria* before injection. Both complete and incomplete adjuvants have been used.

(iii) The number of animals used in injections. In general, at least three animals should be used for each immunogen to be tested.

(iv) The amount of injected antigen. This depends on the immunogenicity of the antigen/hapten and the amount of hapten conjugated to each molecule of protein carrier, as well as the toxicity of the immunogen.

In early studies, antibodies against aflatoxin were produced in rabbits by injecting a mixture of aflatoxin–BSA conjugate in complete Freund's adjuvant using the subcutaneous and/or intramuscular route. Subsequent booster injections were usually made by the intramuscular route in the thigh and incomplete adjuvant was used. In the recent studies, mice are used for monoclonal antibody preparation (Chu, 1984*b*).

2.2.2 Purification of antibody

An antibody is a member of the family of mildly glycosylated proteins called immunoglobulins which can specifically combine with an antigen. One antibody molecule has two 'paratopes' and each of these paratopes is specific for the same epitope (antigenic determinant). Each immunoglobulin (Ig) is composed of equal molar amounts of heavy and light polypeptide chains (Fig. 4). In humans and many common animals, two light chain isotypes, i.e. kappa (κ) and lambda (λ), and five heavy chain isotypes, i.e. IgG, IgM, IgA, IgD and IgE, have been identified (Butler, 1980). IgA and IgM exist in dimeric and pentameric form, respectively. Although IgA, IgG and IgM are involved in the immunochemical methods to be described later, IgG, which represents about 75% of the total immunoglobulin, is the major immunoglobulin used in the immunoassays.

In the normal serum, about 10% of the antiserum proteins are immunoglobulins of which usually 1–10% are antibodies specific to the injected antigen. Enrichment for desired antibodies can be achieved by the purification of immunoglobulins. The most common procedure for the separation of IgG from other proteins in serum is based on selective precipitation with ammonium sulfate. However, a combination of ammonium precipitation with chromatography on DEAE cellulose is preferable. To purify antibody for a specific ligand, immunoaffinity chromatography as mentioned earlier should be used (see Section 1.2).

2.3 Production of Monoclonal Antibody

Animals can produce several different types of immunoglobulins against a given epitope. These immunoglobulins are synthesized by several cell types of lymphocytes. Such immune serum contains polyclonal antibodies which generally are not mono-specific to the given antigen/hapten. Thus, the collected purified antibodies can also

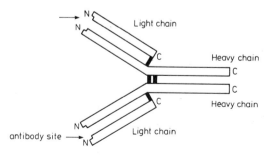

Fig. 4. Basic structure of an immunoglobulin (after Roitt *et al.*, 1985).

react or 'cross-react' with some structurally-related haptens. Consequently, it is important to know the cross-reactivity of such antisera before using them for immunoassay. For practical application, affinity purification of polyclonal antisera may be necessary.

Since it is often impossible to discard the undesired antibodies without losing a substantial part of the desired antibodies, a purification step may result in a lower titer. In addition, it is very difficult to obtain antisera with identical properties from different animals. Even from the same animal, the antisera successively collected at different times can have different properties. These limitations of polyclonal antibodies justify attempts to produce monoclonal antibodies which have identical physical, biochemical and immunological characteristics.

As shown in Fig. 5, the following steps are involved in the production of monoclonal antibodies. The antibody-producing lymphocytes are isolated by removing spleens of mice which have been immunized with

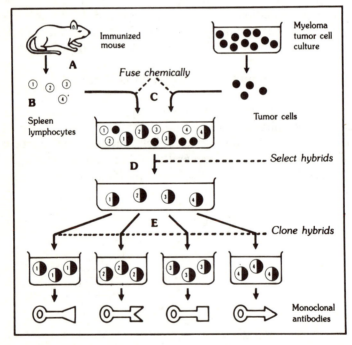

Fig. 5. Illustration of steps involved in the production of monoclonal antibodies (after Hahn & Felsburg, 1983).

the immunogen. These cells are then fused with myeloma cells in the presence of polyethylene glycol. The hybrid cells obtained after fusion are selected in a culture medium that only permits the growth of these hybrids. The supernatant fluids of each of these cells in the tissue culture are screened for the specific antibody produced. The hybrid cells are cloned to separate each individual type of antibody-producing hybridoma. Subsequent subcloning results in a clone which only produces a specific monoclonal antibody. These cells can maintain *in vitro* over long periods of time and can generate large quantities of identical antibody. They are immortal. Monoclonal antibodies for a number of mycotoxins, including AFB_1 and AFM_1, have been obtained in recent years.

2.4 Antibody Specificity

2.4.1 Polyclonal antibodies against aflatoxin M_1

Polyclonal antibodies against AFM_1 were obtained when rabbits were immunized with aflatoxin conjugated through the cyclopentenone ring of the molecule, such as $CMO-AFM_1$ (Harder & Chu, 1979). In this case, the antibodies recognized the dihydrofuran moiety of the aflatoxin molecule. They were specific to AFM_1 but cross-reacted with AFB_1 slightly. When conjugates were prepared through the dihydrofuran portion of the aflatoxin molecule, such as in AFB_{2a} (Lau *et al.*, 1981), and AFB_1-Cl_2 (Sizaret *et al.* 1982), the antibodies showed a specificity towards the cyclopentenone ring. In this case, the antibodies reacted not only with AFB_1 but also with other aflatoxin analogs including AFM_1. Since large amounts of AFM_1 are not commercially available for preparing immunogen, this procedure could be an inexpensive way to produce antibodies against AFM_1 by using AFB_1 derivative conjugates (see Table 1).

2.4.2 Monoclonal antibodies against aflatoxin M_1

In the early 1980s, several methods for the preparation of monoclonal antibodies against AFM_1 were reported. In addition, some of the monoclonal antibodies against AFB_1 which have reactivity against AFM_1 could also be used in the immunoassay for AFM_1 (see Table 1).

A monoclonal antibody which could recognize AFB_1 and AFM_1 equally was first reported in 1983 (Lubet *et al.*, 1983). In 1984, Woychik *et al.* (1984) reported the production of an IgG (2b type) monoclonal antibody specific to AFM_1 and cross-reacting slightly with AFB_1. This

TABLE 1

Cross-reactivity of antibody against aflatoxin M_1 with selected aflatoxins

Aflatoxin derivatives[a]	Antibody types[b]	Cross-reactivity[c]					References
		B_1	B_2	G_1	G_2	M_1	
M_1–CMO	P	+/−	−	−	−	+	Harder & Chu (1979)
B_{2a}–HG	P	+	+/−	−	−	+/−	Lau et al. (1981)
B_1–Cl$_2$	P	+	+	?	−	+/−	Sizaret et al. (1982)
B_{2a}–HG	M	+	+	?	?	+	Lubet et al. (1983)
B_1-2,3-epoxide	M	+/−	+	+/−	+/−	+	Groopman et al. (1984)
M_1–CMO	M	+/−	−	+/−	−	+	Woychik et al. (1984)
B_{2a}	P	+	?	−	?	+	Jackman (1985)
M_1–CMO	P	+/−	−	−	−	+	Martlbauer & Terplan (1985)
B_1–CMO	M	+	+/−	+/−	−	−	Candlish et al. (1985)
B_1–CMO	M	+	−	+/−	−	+/−	Kaveri et al. (1987a)

[a] Derivatives used in the preparation of immunogens. The abbreviations used are: CMO, carboxymethyl oxime; HG, hemiglutarate; Cl$_2$, dichloride.

[b] Type of antibodies: P, rabbit polyclonal; M, monoclonal.

[c] Cross-reactivity (or specificity) to aflatoxins: +, +/− and −, represents high (50–100%), medium (10–49%), and low (<10%) cross-reaction (CR), respectively; not mentioned.

difference in specificity between AFB_1 and AFM_1 is not surprising since Woychik *et al.* (1984) used AFM_1 whereas Lubet *et al.* (1983) employed AFB_{2a} as the hapten in the immunogen preparation. A monoclonal antibody (IgM-type) which interacted similarly with AFB_1, AFB_2 and AFM_1 was obtained by Groopman *et al.* (1984) when they immunized mice with an immunogen prepared by conjugation of AFB_1 to bovine gamma-globulin. More recently, Kaveri *et al.* (1987*a*) obtained a monoclonal antibody, IgG (2a, κ-type), which has specificity for both AFB_1 and AFM_1. Kornfeld *et al.* (1987) recently obtained a monoclonal antibody specific to AFB_1, AFB_2, AFG_1 and AFM_1.

3 IMMUNOCHEMICAL PROCEDURES FOR DETERMINATION OF AFLATOXIN M_1

3.1 Radioimmunoassay (RIA)

3.1.1 Principle
The use of radio isotopes such as ^3H, ^{14}C, and ^{125}I labelled haptens, antigens and antibody permits detection of trace amounts of many substances. In a typical RIA, three steps are involved:

(1) Incubation. A constant amount of antibody is incubated together or sequentially with reference/standard unlabelled antigen (analyte) or the unknown sample (containing analyte) and with a constant amount of labelled antigen. The labelled antigen, also called a radio ligand tracer or marker, competes with the unlabelled antigen for binding with the antibodies. After a certain time, when equilibrium is reached, the solution would contain both free antigen as well as antigen bound to the antibody (see Fig. 6). Since the goal of the analysis is to distinguish the proportion of the labelled antigen/analyte in the free and in the antibody-bound forms, a separation between them is then carried out.

(2) Separation. Because of the large differences in molecular weight between the free antigen (a hapten in this case) and the antibody-bound antigen complex, separation can be achieved by physicochemical means. Ammonium sulfate precipitation and charcoal adsorption methods are most commonly used.

(3) Radioactivity measurement. The radioactivity of the complex is measured either in a liquid scintillation counter for ^3H or ^{14}C labelled antigens (haptens) or in a gamma counter for ^{125}I labelled antigens.

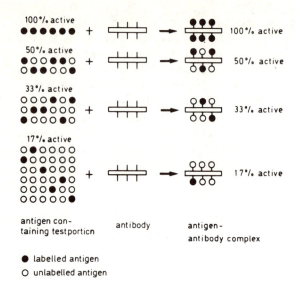

Fig. 6. Principles of radioimmunoassay. The open circles indicate unlabelled antigen (or toxin) and the closed circles indicate labelled antigen (or marker ligand) (after Van Egmond, 1984).

Because the amount of radioactive marker in the system is constant, the ratio of the relative binding of the labelled and unlabelled antigen with the antibody depends on the relative concentration of the labelled and unlabelled antigen in the test solution. Thus, the higher the concentration of the unlabelled antigen present in the sample, the lower the amount of radioactivity appearing in the antigen–antibody complex. The radio-activity is inversely related to the amount of unlabelled antigen present in step 1 as shown in Fig. 6 (Van Egmond, 1984).

3.1.2 Determination of aflatoxin M_1 in milk

For RIA of AFM_1, the protocol often used includes simultaneous incubation of a test solution of an unknown amount of AFM_1 (or a standard solution of a known amount of AFM_1 in phosphate buffer) with a constant amount of labelled aflatoxin and specific antibody. Free and bound labelled aflatoxin are then separated, and the radioactivity is determined. The AFM_1 concentration of the test solution is determined by comparing the results to a standard curve which is established by plotting the ratio of bound and the initial (total) amount of labelled aflatoxin multiplied by 100 (% binding), versus the AFM_1 standard concentration (Chu, 1984*a*).

Tritiated aflatoxin B_1 (^{3}H-AFB_1) rather than ^{3}H-AFM_1 is usually used as the radio ligand tracer due to its commercial availability at low cost (Pestka *et al.*, 1981; Sizaret *et al.*, 1982; Qian *et al.*, 1984). Sizaret *et al.* (1982) applied the test only to aflatoxin residues in urine. Pestka *et al.* (1981) obtained a standard curve for RIA of AFM_1 at concentrations ranging from 5 to 50 ng per assay; the detection limit for AFM_1 in whole milk was 0·5 ng/ml. Using a more complicated cleanup treatment, Qian *et al.* (1984) could detect as little as 30 pg AFM_1 in 12 ml of milk in the RIA; the standard curve for this analysis is shown in Fig. 7. The detection limit of RIA of AFM_1 improved considerably when Rauch *et al.* (1987) most recently used ^{125}I-tyramine–AFB_1 as the labelled marker. The detection limit of this method was 4 pg per assay and the concentration causing a 50% inhibition of binding of labelled marker to the antibody was 100 pg per assay (Rauch *et al.*, 1987).

3.2 Enzyme Immunoassay (EIA)

3.2.1 Principle
There are two types of EIA. The homogeneous EIA in which no separation of the reagents is performed: and the heterogeneous EIA in which a separation between the immuno complex and the unreacted material is performed. One of the most common heterogeneous EIAs, known as enzyme-linked immunosorbent assay (ELISA), is generally used for AFM_1 determination. In this assay, one reagent (antibody or antigen) is immobilized on a solid phase e.g. plastic tubes, beads,

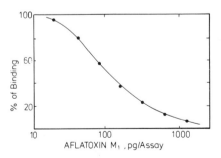

Fig. 7. Standard curve for radioimmunoassay for aflatoxin M_1 in milk. Concentrations of aflatoxin M_1 are in log scale. (Data obtained from Qian *et al.*, 1984.)

microplates. The test solution or the toxin standard and an enzyme labelled antigen or labelled antibody are then added. The labelling enzymes generally are alkaline phosphatase (AP) or horseradish peroxidase (HRP). After incubation and washing to eliminate free and unreacted material, the amount of enzyme bound to the solid-phase is measured by adding chromogenic substrates such as *para*-nitro-phenyl-phosphate (pNPP) for AP enzyme, or *ortho*-phenylene-diamine (oPD), 2, 2 azino-di-3-ethyl-benzenethiazoline-sulfonate (ABTS), tetra-methyl-benzidine (TMB) for HRP enzyme. The resulting color is then measured spectrophotometrically or by visual comparison with standards (Chu, 1984*a*). Again, like RIA, the amount of analyte bound to the immobilized antibody or immobilized antigen is determined by competition with enzyme labelled antigen or labelled antibody.

Two types of ELISA, i.e. *direct* and *indirect*, have been applied in aflatoxin assays. As shown in Fig. 8, in the direct ELISA, the specific antibody is coated on a solid phase and an aflatoxin–enzyme conjugate is used as the marker. However, in the indirect ELISA, as shown in Fig. 9, a polypeptide–aflatoxin conjugate is coated on the solid phase and a second antibody to which an enzyme has been conjugated is used. In the indirect assay, if the first anti-AFM$_1$ antibody is collected from rabbits, the second antibody should be an anti-rabbit immunoglobulin. Since many types of second antibody–enzyme conjugates are commercially available, the analysts do not have to prepare their own enzyme conjugate in the indirect ELISA.

Several steps are involved in both direct and indirect ELISA. The first step is coating either antigens or antibodies in the wells of polystyrene plates (microtiter plates). Each step is separated by careful washing with saline solutions containing low amounts of detergents in order to prevent non-specific adsorptions. Blanks are carried out simultaneously by using dilution buffers, and in some cases pre-immune sera (collected prior to immunization) are used.

3.2.2 Determination of aflatoxin M$_1$ in dairy products

Direct competitive ELISA (see Fig. 8) was used in most of the works reported in the early 1980s. In these assays, specific rabbit polyclonal antibodies are first coated to a microtiter plate. The test solution or AFM$_1$ standard is generally incubated simultaneously with the aflatoxin–enzyme conjugate. The systems involving HRP (Pestka *et al.*, 1981; Hu *et al.*, 1983; Frémy & Chu, 1984; Martlbauer & Terplan, 1985) or AP (Jackman, 1985) as enzyme and ABTS or pNPP respectively as

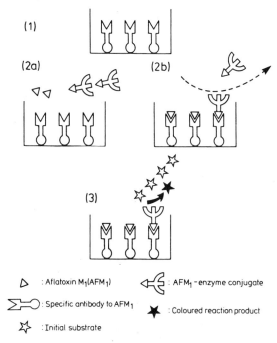

Fig. 8. Schematic diagram for the direct competitive ELISA for aflatoxin M_1 determination. Specific antibody against AFM_1 is first coated on the microtiter plate (1) which is then incubated with AFM_1–enzyme conjugate in the presence of sample or standard AFM_1 (2a, initial incubation). After washing to remove the unreacted toxin or toxin–conjugate (2b), a chromogenic substrate is then added to each well. After incubation (3), the substrate changes color due to the formation of reaction product.

substrate have been mostly used. As shown in Fig. 8, the higher the AFM_1 concentration in the test solution, the less the toxin–enzyme conjugate binds to the solid-phase antibody. Consequently, a lower concentration of reaction product, which resulted in a lower color reaction, was formed. Like RIA, the toxin concentration in the test solution is then determined from a standard curve, as shown in Fig. 10. The detection limits for AFM_1, together with information about the milk product, the type of contamination and cleanup procedures of some published ELISA methods for AFM_1 are presented in Table 2. Milk samples can be used directly in the ELISA (Pestka *et al.*, 1981). However, the detection limits can be drastically reduced to around

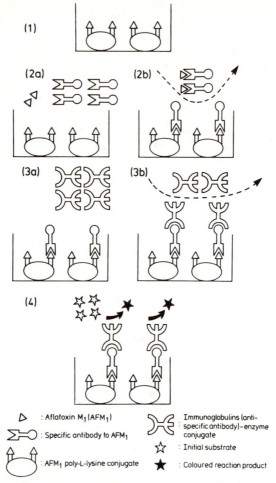

Fig. 9. Schematic diagram for the indirect competitive ELISA for aflatoxin M_1 determination using a secondary antibody–enzyme conjugate as the marker. AFM_1–polylysine conjugate is first coated on the microtiter plate (1) and then incubated with specific antibody (rabbit IgG) against AFM_1 together with either sample or standard AFM_1 (2a) solution. After washing to remove the excess toxin and anti-AFM_1 antibody (2b), a second antibody (anti-specific antibody such as goat anti-rabbit IgG) conjugated with an enzyme is added and incubated again for an appropriate time (3a). After washing (3b), a chromogenic substrate is then added to each well. After incubation (4), the substrate changes color due to the formation of reaction product.

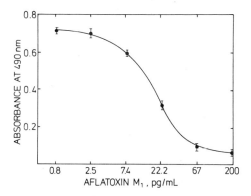

Fig. 10. Standard curve for direct competitive ELISA for aflatoxin M_1 in milk. Concentrations of aflatoxin M_1 are in log scale. (Data obtained from Martlbauer & Terplan, 1985.)

5–15 ng/litre if milk samples are subjected to a simple cleanup treatment with reversed phase micro-columns (Frémy & Chu, 1984; Martlbauer & Terplan, 1985). The direct ELISA has been also successfully applied to yogurt and cheese samples (Frémy & Chu, 1984).

Because AFM_1–enzyme conjugates are not very stable over a long

TABLE 2
Detection limits for ELISA for aflatoxin M_1 in dairy products

Product	Contamination	Cleanup	Detection limits	Reference
Milk	Spiked	None	0.25^a	Pestka *et al.* (1981)
Milk	Spiked	RP-col[b]	0.015	Hu *et al.* (1983)
Milk	Spiked	RP-col	0.005	Frémy & Chu (1984)
Milk	Natural	RP-col	0.01	Frémy & Chu (1984)
Milk	Natural	RP-col	0.10	Jackman (1985)
Milk	Natural	RP-col	0.005	Martlbauer & Terplan (1985)
Yogurt	Natural	NP-col[c]	0.01	Frémy & Chu (1984)
Cheese	Natural	NP-col	0.05	Frémy & Chu (1984)
Milk	Spiked	NP-col	0.03	Kornfeld *et al.* (1987)
Milk	Spiked	RP-col	0.05	Kaveri *et al.* (1987*b*)

[a]Concentrations: ng/ml for milk or ng/g for yogurt and cheese.
[b]RP-col: C_{18} reversed-phase cartridge.
[c]NP-col: Si-60 normal phase cartridge.

time, an indirect ELISA (see Fig. 9) could be used to overcome this problem. In such assays, an aflatoxin–polypeptide conjugate is first coated on the microplate. Conjugation of AFM_1 to polypeptide such as poly-L-lysine (PLL) is similar to the method for the preparation of toxin–protein conjugates (see Section 2.1). Specific antibodies are incubated in the presence or absence of AFM_1 in the wells of a microplate which have been coated with the aflatoxin–polypeptide conjugate. Thus, the antibody is either bound to the aflatoxin–PLL in the solid-phase or bound with the toxin in the solution. The amount of first antibody (such as rabbit IgG) bound to aflatoxin–PLL is determined by an anti-antibody–enzyme complex (such as goat anti-rabbit-IgG-HRP) and subsequent reaction with an adequate substrate (see Fig. 9). Many second antibody–enzyme complexes are commercially available and are stable for a long time. The effectiveness of this technique has been successfully tested for several mycotoxins (Chu, 1984a) including AFM_1 (Kaveri et al., 1987b). An alternative to the indirect ELISA is using the specific antibody conjugated to an enzyme (see Fig. 11) as the marker. In this case, the aflatoxin–polypeptide conjugate is coated on the microplate which is then incubated with the specific antibody-enzyme conjugate in the presence or absence of AFM_1. Thus, it can also avoid use of AFM_1–enzyme conjugates. This type of indirect competitive ELISA has been recently applied to AFM_1 determination in milk (Kornfeld et al., 1987).

Using monoclonal antibodies (see Section 4.2) and the indirect ELISA, the detection limits in milk were found to be 50 ng/litre by Kaveri et al. (1987b) and 30 ng/litre by Kornfeld et al. (1987) (see Table 2).

Further improvement of EIA sensitivity could be achieved by the use of biotin–avidin or biotin–streptavidin systems. Antibodies or enzymes can be efficiently 'biotinylated' without significant loss of biological activity. Avidin, an egg-white protein, has a high affinity for biotin and can be conjugated with a variety of labels per molecule. The signal can be also enhanced by the use of streptavidin containing several labels per molecule. The combination of these factors could provide sufficient amplification to detect very low levels of antigen/hapten.

3.3 Immunoaffinity Chromatography (IAC)

As mentioned earlier (see Section 1.2), immunoaffinity chromatography can be used to retain the antigen/hapten or antibody. An example of

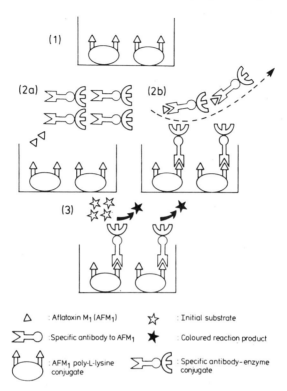

Fig. 11. Schematic diagram for the indirect competitive ELISA for aflatoxin M_1 determination using a specific antibody–enzyme conjugate as the marker. AFM_1–polylysine conjugate is first coated on the microtiter plate (1) and then incubated with specific antibody against AFM_1 conjugated to an enzyme together with either sample or standard AFM_1 (2a) solution. After washing to remove the excess toxin and anti-AFM_1 antibody–enzyme conjugate (2b), a chromogenic substrate is then added to each well. After incubation (3), the substrate changes color due to the formation of reaction product.

using antibody as the affinity reagent (and thus, retain antigen/hapten) is shown in Fig. 12. Several steps are involved in immunoaffinity chromatography. Specific antibody is first immobilized on a solid support, e.g. Sepharose® or Ultrogel®, which is then packed in a small column. The test solution is passed through the column. Thus, antigen/ hapten is adsorbed on the solid-phase gel by binding with the specific antibody. After washing with a buffer solution at neutral pH to remove impurities, the antigen/hapten is desorbed by using an appropriate

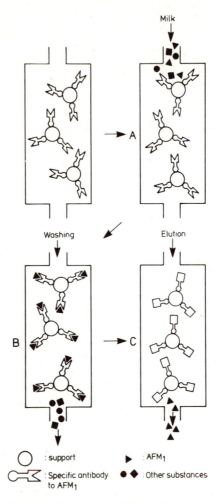

Fig. 12. Schematic diagram for immunoaffinity chromatography for concentration and purification of aflatoxin M_1. Sample containing AFM_1 is first loaded to the affinity gel column containing antibody against AFM_1 (A). After washing to remove impurities (B), the AFM_1 is eluted from the column with methanol (C).

buffer solution at a lower pH, or by using an alcoholic solution which is then subjected to other quantitation techniques such as immuno-chemical or physicochemical methods.

3.3.1 Immunoaffinity chromatography and RIA

Immunoaffinity chromatography has been used in combination with RIA for determining aflatoxin in very complex matrices as early as in the 1970s (Sun & Chu, 1977). After the typical steps in immunoaffinity chromatography (see Section 3.3), the radioactivity in the eluted materials (unadsorbed free antigen) or in the dissolved fraction (bound form) is then measured. Like the regular RIA, the amount of radioactivity in the eluate is inversely related to the amount of aflatoxin in the reference or in the unknown sample solutions. This technique got wider application in the 1980s, when monoclonal antibodies against aflatoxin became available. Monoclonal antibody specific for AFB_1, AFB_2, AFM_1 and major aflatoxin–DNA adducts (Groopman *et al.*, 1984), was covalently bound to a Sepharose-4B®. Aflatoxins added *in vitro* to phosphate buffer, human urine, human serum and human milk (0·1 ng/ 10 ml) were quantitatively recovered by applying the mixture to the immunoaffinity column and by measuring the recovery of the ^3H-AFB_1 radioactivity (Groopman *et al.*, 1985) added to the system. Also, the monoclonal antibody obtained by Kaveri *et al.* (1987*a*) previously described was covalently bound to Ultrogel AcA 22® support. Tritiated AFB_1 added *in vitro* to bovine milk at 0·1 ng/ml was quantitatively recovered in the eluted fraction (Kaveri *et al.*, 1987*a*). In both cases, these authors successfully used and regenerated their respective affinity chromatography columns several times (Groopman *et al.*, 1985; Kaveri *et al.*, 1987*a*).

3.3.2 Immunoaffinity chromatography and HPLC

Immunoaffinity was also used in combination with HPLC for determining AFM_1 in milk. In this case, the affinity column was included in the extraction and cleanup steps (Mortimer *et al.*, 1987). Milk was first passed through an 'Easi-extract®' column, an affinity column containing anti-AFM_1 antibody, either manually or by using a syringe pump. The column was then washed with water and eluted with acetonitrile. The acetonitrile eluate was evaporated to near dryness; then 100–400 µl water/acetonitrile were added. This solution was then injected in the HPLC system for separation and fluorescence detection. Recovery of AFM_1 from skim milk spiked at 0·1 µg/litre averaged 85·7%.

Because reconstituted dried milks contain some particles, volumes in excess of 100 ml caused unacceptable build-up of back-pressure in the column. However, 25 or 40 ml of reconstituted milk sample can be loaded to the column. More than 10 ml of whole milk blocked the column. After centrifugation to remove the fat, samples of at least 1000 ml could be extracted on one column. Large volumes of skimmed milk can be analyzed directly. In all cases sufficient volumes could be analyzed to give detection limits below 10 ng/litre (10 ppt). Because HPLC chromatograms are significantly cleaner after treating samples by this method than by other methods. Mortimer *et al.* (1987) estimated that a sample size of 1000 ml should provide a detection limit for AFM_1 in milk about 50 pg/kg (0·050 ppt). A similar approach for the analysis of AFM_1 in human urine was reported by Wu *et al.* (1983).

4 VALIDATION AND COMPARISON OF IMMUNOCHEMICAL PROCEDURES

Aflatoxins have defined spectral properties which permit direct chemical analysis after separation of different structurally related compounds. Only trace amounts of toxins are present in complex matrices, e.g. AFM_1 in milk and dairy products. Therefore, extensive cleanup is necessary to remove large amounts of impurities before any determination using chromatographic techniques such as high pressure liquid chromatography (HPLC) and thin layer chromatography (TLC) can be done. Despite the improvements in chromatographic procedures made over the years, intensive cleanup for the sample remains necessary to reach the required low detection limits for AFM_1 in dairy products. This makes these procedures rather time-consuming. In addition, expensive instrumentation (except for TLC, with visual estimation of toxin) is necessary (Chu, 1984*b*). In principle, immunoassays should have some advantages over chromatographic methods. Comparative studies for the determination of AFM_1 in milk by immunoassays with other methods, e.g. RIA vs HPLC (Qian *et al.*, 1984), ELISA vs TLC (Jackman, 1985; Martlbauer & Terplan, 1985), and ELISA vs HPLC (Frémy & Chu, 1984), revealed that immunoassays were indeed more rapid and easy to perform, and lower limits of detection could be achieved.

Comparisons between RIA and ELISA for the determination of AFM_1 in milk have also been made. Because of its simplicity, Pestka *et al.,* (1981) concluded that ELISA was the preferred method. Taking into account the disadvantages of using radioactive substances such as the limited shelf-life of certain reagents, the legislative inconveniences relating to supply, use and disposal of isotopes, and together with the 'anti-nuclear' social issues, it is to be expected that ELISA is growing in importance as an assay technique (Chu, 1984*a, b*). In general, RIA will probably decrease in its application for routine analysis (Voller & Bidwell, 1983).

Although there are many advantages of using ELISA for the analysis of AFM_1 in dairy products, very low levels of AFM_1, i.e. less than 0·1 µg/kg, still have not been measurable via direct loading of milk into microtiter plates. Alternative strategies, involving some form of sample cleanup lose one of the main advantages of ELISA: its simplicity and consequent sample throughput. The combination of affinity column cleanup with HPLC determination recently reported by Mortimer *et al.* (1987) seems to give excellent results for recovery, detectability and sample throughput. These authors also compared their procedure with the procedure of Takeda (1984) which gave the best results in a previous comparison of six published methods for AFM_1 (Shepherd *et al.*, 1986). The affinity column permitted a detection limit lower by a factor of 100 and had a sample throughput 50–100% greater when analyzing equal volumes of milk. Although affinity columns are relatively expensive and cost three times more than a C_{18} reversed-phase extraction cartridge, the overall cost per analysis is lower because of the reduction in the labor element (Mortimer *et al.*, 1987).

In conclusion, due to their simplicity and specificity, the immunochemical methods represent a great potential for a rapid determination of aflatoxin residues in complex matrices such as dairy products. Recent development of immunoaffinity column cleanup in combination with HPLC method or immunoassays such as ELISA appears to be very promising for routine AFM_1 analysis. Radioimmunoassays are less and less used. In some immunoassays, high variation coefficients were noticed (Chu, 1984*a*). Thus, improvements of immunoassay procedures and extensive collaborative studies should be made to validate the performance characteristics of immunoassays such as reproducibility and repeatability, accuracy, sensitivity, detectability, specificity and selectivity. Although such additional research efforts are still necessary, the use of immunochemical methods either alone or in combination with physicochemical methods, should be considered next as a potentially powerful tool for sensitive and rapid AFM_1 determination. At the present time, immunoassays for drugs and other biologically active substances are widely used; the whole diagnostic industry is expanding. Recent commercial availability of reagents and test kits for aflatoxin determination should widen the scope of use of this new method in mycotoxin methodology in the next few years.

5 SUMMARY

Immunochemical methods have been recently used for determination of biologically active substances including aflatoxin M_1 (AFM_1). Since aflatoxins are low molecular weight substances, they are non-immunogenic and are unable to elicit antibodies when they are injected into an animal. Aflatoxins must first be conjugated to a protein before immunization. Both polyclonal and monoclonal antibodies have been produced and characterized for immunoassays for AFM_1. In the radioimmunoassay (RIA), either tritiated aflatoxin B_1 (T-AFB_1) or iodinated-tyramine-AFB_1 (I-AFB) was used as the marker ligand with a detection limit for AFM_1 in liquid raw milk in the RIA of 3 ng/litre. In the enzyme-linked immunosorbent assay (ELISA), either specific antibodies or reference aflatoxin was immobilized on a solid phase and the enzyme labelled aflatoxin or antibodies was used as the marker. The detection limits for ELISA for AFM_1 in milk could reach to 5–10 ng AFM_1/litre. Antibody against AFM_1 has also been used to prepare an immunoaffinity column which was subsequently used as a cleanup column to concentrate AFM_1 from milk or urine samples and to remove impurities before further analyses. This purification technique in combination with HPLC permitted detection as low as 0·05 ng AFM_1/litre of skim milk. Details for the production and characterization of antibodies against AFM_1, the cross-reactivity of AFM_1 antibodies against different aflatoxin-related compounds, immunoassay protocols and the application of immunoassays for AFM_1 in different commodities are reviewed in this chapter.

ACKNOWLEDGMENT

The authors wish to thank Ms. Susan Schubring for her help in the preparation of this manuscript.

REFERENCES

Butler, J. E. (1980). Antibody–antigen and antibody–hapten reactions. In *Enzyme-immunoassay*, ed. E. T. Maggio. CRC Press, Boca Raton, Florida, pp. 5–52.

Candlish, A. A. G., Stimson, W. H. & Smith, J. E. (1985). A monoclonal antibody to aflatoxin B_1: detection of the mycotoxin by enzyme immunoassay. *Lett. Appl. Microbiol.,* **1**, 57–61.

Chu, F. S. (1984*a*). Immunochemical studies on mycotoxins. In *Toxigenic Fungi,* ed. H. Kurata, & Y. Ueno, Kodansha Ltd, Tokyo, Japan, pp. 234–44.

Chu, F. S. (1984*b*). Immunoassays for analysis of mycotoxins. *J. Food Prot.,* **47**, 562–9.

Chu, F. S. & Ueno, Y. (1977). Production of antibody against aflatoxin B_1. *Appl. Environ. Microbiol.,* **33**, 1125–8.

Daussant, J. & Burreau, D. (1984). Immunochemical methods in food analysis. In *Developments in Food Analysis Techniques, III,* ed. R. D. King. Applied Science Publishers, London, pp. 175–210.

Fan, T. S., Zhang, G. S. & Chu, F. S. (1984). Production and characterization of antibody against aflatoxin Q_1. *Appl. Environ. Microbiol.,* **47**, 526–32.

Frémy, J. M. & Chu, F. S. (1984). Direct ELISA for determining AFM_1 at picogram levels in dairy products. *J. Assoc. Off. Anal. Chem.,* **67**, 1098–101.

Groopman, J. D., Trudel, L. J., Donahue, P. R., Rothstein, A. N. & Wogan, G. N. (1984). High affinity monoclonal antibodies and their applications to solid-phase immunoassays. *Proc. Natl. Acad. Sci., USA,* **81**, 7728–31.

Groopman, J. D., Donahue, P. R., Zhu, J., Chen, J. & Wogan, G. N. (1985). Aflatoxin metabolism in humans: detection of metabolites and nucleic acids adducts in urine by affinity chromatography. *Proc. Natl. Acad. Sci., USA,* **82**, 6492–6.

Hahn, E. C. & Felsburg, P. J. (1983). Animal immunology: Unraveling the mystery. *Illinois Research,* **25**, 17–19.

Harder, W. O. & Chu, F. S. (1979). Production and characterization of antibody against aflatoxin M_1. *Experientia,* **35**, 1104–6.

Hu, W. J., Woychik, N. & Chu, F. S. (1983). ELISA of picogram quantities of AFM_1 in urine and milk. *J. Food Prot.,* **47**, 126–7.

Jackman, R. (1985). Determination of aflatoxins by ELISA with special reference to aflatoxin M_1 in raw milk. *J. Sci. Food Agric.,* **36**, 685–98.

Kaveri, S. V., Frémy, J. M., Lapeyre, C. & Strosberg, A. D. (1987*a*). Immuno-detection and immunopurification of aflatoxins using a high affinity monoclonal antibody to AFB_1. *Lett. Applied Immunol.,* **4**, 71–5.

Kaveri, S. V., Lapeyre, C., Reboul, S. & Frémy, J. M. (1987*b*). Aflatoxin determination in feed and food samples by indirect ELISA using mono-clonal antibody. In *Abstracts of the Hybridoma Conference*, Paris, France, 25–27 November 1987, Innobio Association, Paris, France, p. 25.

Kornfeld, S., Hurpet, D. & Bonneau, M. (1987). Determination of aflatoxins in foodstuffs and milk by ELISA. In *Abstracts of the Hybridoma Conference*, Paris, France, 25–27 November 1987, Innobio Association, Paris, France, p. 29.

Kurstak, E. (1985). Progress in enzyme immunoassays: production of reagents, experimental design, and interpretation. *Bull. World Health Org.,* **63**, 793–811.

Lau, H. P., Gaur, P. K. & Chu, F. S. (1981). Aflatoxin B_{2a} — hemiglutarate and antibody production. *J. Food Safety,* **3**, 1–13.

Lubet, M., Olson, D. F., Yang, G., Ting, R. & Steuer, A. (1983). Use of a monoclonal antibody to detect aflatoxin B_1 and M_1 in enzyme immunoassay. *Abstr. Ann. Int. Meet. Assoc. Off. Chem.,* **177**, 71–2.

Martlbauer, E. & Terplan, G. (1985). Highly specific heterologous enzyme immunoassay for aflatoxin M_1 in milk and milkpowder. *Arch. für Lebensmit.,* **36**, 53–5.

Mortimer, D. N., Gilbert, J. & Shepherd, M. J. (1987). Rapid and highly sensitive analysis of aflatoxin M_1 in liquid and powdered milks using an affinity column cleanup. *J. Chromatogr.,* **407**, 393–8.

Oakley, C. L. (1970). Immunology of bacterial protein toxins. In *Microbial Toxins, Vol. 1,* ed. S. J. Kadis & T. C. Montie. Academic Press, New York.

Ouchterlony, Ö. (1949). Antigen–antibody reaction in gel. *Ark. Kem. Mineral. Ged.,* **26**, 1–9.

Pestka, J. J., Li, Y., Harder, W. O. & Chu, F. S. (1981). Comparison of RIA and ELISA for determining aflatoxin M_1 in milk. *J. Assoc. Off. Anal. Chem.,* **64**, 294–301.

Qian, G. S., Yasei, P. & Yang, G. (1984). Rapid extraction and detection of aflatoxin M_1 in cow milk by HPLC and RIA. *Anal. Chem.,* **56**, 2079–80.

Rauch, P., Fukal, L., Prosek, J., Bresina, P. & Kas, J. (1987). Radioimmunoassay of aflatoxin M_1. *J. Radioanal. Nucl. Chem. Lett.,* **117**, 163–9.

Roitt, I., Brostoff, J. & Male, D. (1985). *Immunology.* Gower Medical Pub. Ltd, London, UK.

Sizaret, P. & Malaveille, C. (1983). Preparation of aflatoxin B_1–BSA conjugate with high hapten/carrier molar ratio. *J. Immunol. Methods,* **63**, 159–62.

Sizaret, P. & Malaveille, C., Montesano, R. & Frayssinet, C. (1982). Detection of aflatoxins and related metabolites by RIA. *J. Natl Cancer Inst.,* **69**, 1375–80.

Shepherd, M. J., Holmes, M. & Gilbert, J. (1986). Comparison and critical evaluation of six published extraction and cleanup procedures for aflatoxin M_1 in liquid milk. *J. Chromatogr.,* **368**, 305–15.

Sun, P. & Chu, F. S. (1977). A simple solid-phase radioimmunoassay for aflatoxin B_1. *J. Food Safety,* **1**, 67–75.

Takeda, N. (1984). Determination of aflatoxin M_1 in milk by reversed-phase high-performance liquid chromatography. *J. Chromatogr.,* **288**, 484–8.

Van Egmond, H. P. (1984). Determination of mycotoxins. In *Developments in Food Analysis Techniques III,* ed. R. D. King. Applied Science Publishers, London, UK, pp. 99–144.

Voller, A. & Bidwell, D. E. (1983). Non-isotopic immunoassays with special reference to labelled reagent methods. In *Non-isotopic Immunoassays and their Applications,* ed. G. P. Talwar. Vikas Publishing House, New Delhi, India, pp. 3–10.

Woychik, N., Hindsill, R. D. & Chu, F. S. (1984). Production and characterization of monoclonal antibodies against aflatoxin M_1. *Appl. Environ. Microbiol.,* **48**, 1096–9.

Wu, S., Yang, G. & Sun, T. (1983). Studies on immuno-concentration and immunoassay of aflatoxins. *Clin. J. Oncol.,* **5**, 81–4.

Yalow, R. S. (1978). Radioimmunoassay: a probe for fine structure of biological systems. *Science,* **200**, 1236–40.

Chapter 5

Stability and Degradation of Aflatoxin M₁

A. E. Yousef & E. H. Marth

Department of Food Science and The Food Research Institute, University of Wisconsin-Madison, USA

1. Introduction.. 127

2. Fate of Aflatoxin M₁ During Processing of Dairy Products................. 128
    2.1 Processes that do not separate milk components
    2.2 Processes that separate milk components
2.3 Possible sources of variability in results

3. Degrading Aflatoxin M₁ in Milk by Special Treatments........... 144
    3.1 Chemical treatments
    3.2 Physical treatments
    3.3 Chemistry of degradation of Aflatoxin M₁

4. Conclusion and a Look to the Future............................ 156

5. Summary.. 156

1 INTRODUCTION

When the feed of lactating animals contains sufficient aflatoxin B_1 (AFB_1), milk from these animals will contain aflatoxin M_1 (AFM_1). Presence of this toxic and possibly carcinogenic compound in milk is a potential threat to the health of consumers of dairy products. In many instances, detection of AFM_1 in milk results in disposal of the milk, causing economic losses for the dairy industry.

The most effective means to produce AFM_1-free dairy products is to eliminate AFB_1 from the feed of lactating animals. Moldy corn (*Zea mays* L.) and the cake produced from moldy cottonseed can be important sources of aflatoxin in the diet of lactating animals. Several methods to detoxify aflatoxin-contaminated feedstuffs have been studied. Treatment with ammonia, however, appears to be the most useful method and has been used on a large scale to detoxify contaminated corn and cottonseed. When contaminated feed is treated with aqueous or gaseous ammonia and heat, aflatoxin can virtually be eliminated from these products (Anonymous, 1979). Since treatment with ammonia may decrease the quality and nutritive value of feed, treatment with bisulphite has been suggested as an alternative method (Moerck *et al.*, 1980; Hagler *et al.*, 1982). One of the suggested procedures involves soaking corn kernels in 10% sodium bisulphite solution for 72 h, removing the solution and incubating the corn in sealed plastic bags at 50°C for 21 days.

In spite of attempts to eliminate aflatoxin from feed and regulatory/ control measures taken by many countries, production of aflatoxin-free milk is not always achieved. Consequently, many studies were done to explore the possibility of destabilizing the toxin during processing of milk, thus precluding the necessity to dispose of the AFM_1-contaminated milk. Other studies focused on treating the milk with physical or chemical agents to degrade or eliminate the toxin. Results of these two types of studies will be addressed in this chapter.

The meaning of a few words and phrases need to be explained as they pertain to the subject matter of this chapter. The phrase 'naturally-contaminated milk' means that lactating animals (mostly cows, but in few instances buffaloes or ewes) were fed an AFB_1-contaminated diet, and then AFM_1-contaminated milk obtained from these animals was used in the study. It may also mean that milk collected from a particular source was found to contain AFM_1, but no AFB_1 was intentionally added to the diet of the animals. 'Artificially-contaminated milk' is used to designate milk which was contaminated through addition of a solution of AFM_1. The word 'degradation' in reference to AFM_1, will be used when chemical modification of the AFM_1 molecule probably occurred. It does not necessarily mean fragmentation of the molecule into smaller entities.

2 FATE OF AFLATOXIN M_1 DURING PROCESSING OF DAIRY PRODUCTS

Milk may be processed into dairy products without separation of its components. Pasteurization of milk and production of whole-milk yogurt are examples of such processes. Usually a physical or biological agent is used in processing these products. Fate of AFM_1 in such products is a question of toxin stability during their processing or storage. Manufacture of other dairy products involves separation of milk components. Examples of such processes include removal of water from milk during the manufacture of dried milk, removal of whey during cheese-making, and separation of milkfat to produce cream. Therefore, partition of AFM_1 between different milk components is the main subject to be addressed in this category of processing procedures. Because heat, acids, or microorganisms may be employed to make these products, stability of toxin during processing can not be overlooked. Each of these approaches to processing of milk will be considered in the following discussion.

2.1 Processes that do not Separate Milk Components

2.1.1 Heat treatments

Treatments used in studies on stability of AFM_1 during heat processing ranged from pasteurization (Allcroft & Carnaghan, 1962) to heating milk directly on a fire for 3–4 h (Patel *et al.*, 1981). Results of early studies are extremely contradictory, as is evident from the following discussion. Allcroft & Carnaghan (1962, 1963) found that neither pasteurization nor roller-drying reduced the toxicity of milk to ducklings. Doubts about the validity of these findings were raised (Purchase *et al.*, 1972) since the method of assay for aflatoxin M in former studies was not sufficiently quantitative to measure small changes in the amount of the toxin that was present (Wogan, 1964). About 10 years after Allcroft & Carnaghan published their studies, Purchase *et al.* (1972) reported the results of their study on stability of aflatoxin M during heat processing of milk. In their study, milk from cows fed AFB_1 was pasteurized, evaporated, roller- and spray-dried, and sterilized. Chemical analysis indicated that heat processing of milk reduced its aflatoxin M content by 32–86%, and that the higher the temperature used, the smaller the amount of aflatoxin that was recovered. Results of a number of studies on stability of AFM_1 during heat processing are currently available (Table 1).

Data summarized in Table 1 suggest that heat processing of dairy products may or may not change the amount of AFM_1 in these products. Reported changes caused by heat processing varied from substantial decreases to an increase in the amount of the toxin in the dairy product. Variability in these data may have resulted from one or more of the factors listed later in this chapter (see Section 2.3). In spite of the variability of research results just indicated, most of the studies indicate

TABLE 1

Results of studies, arranged in chronological order, on stability of aflatoxin M_1 as affected by heat processing of dairy products

Product	Source of milk contamination	Heat treatment	Changes in amount of AFM_1	Reference
Milk	Natural	Pasteurization (flash or holder)	No change	Allcroft & Carnaghan (1962)
Milk	Natural	Pasteurization	32–64% loss	Purchase et al. (1972)
		Sterilization	81% loss	
		Evaporation	64% loss	
		Roller-drying	61–76% loss	
		Spray-drying	86% loss	
Cheese curd:	Artificial			
Queso Blanco		Curd cooked at 82°C	No change	Stubblefield & Shannon (1974)
Ricotta		Cooking included heating at 100°C	No change	
Milk	Artificial	Pasteurization at 63°C for 30 min	No change	Stoloff et al. (1975)
	Natural	Pasteurization at 77°C for 16 s	No change	

Product	Type	Treatment	Result	Reference
Unripened cheese	Natural	Heated at 80°C for 5 min	No change	Kiermeier & Buchner (1977b)
Milk	Natural or artificial	Temp: 71–120°C Time: 40 s–30 min	12–35% loss	Kiermeier & Mashaley (1977)
Emmental cheese	AFM_1 added to cheese	Melting at 90°C for 3–30 min	No change	Polzhofer (1977)
Milk	Natural	Pasteurization (two methods) Sterilization (two methods)	No change No change	Van Egmond et al. (1977)
Milk	Natural	Direct heating for 3–4 h	No change	Patel et al. (1981)
Cheddar cheese	Natural	Steam injection (90°C for 20 min)	34% increase	Brackett & Marth (1982b)
Milk	Natural or artificial	Heating at 64–100°C for 15–120 min	No obvious change	Wiseman & Marth (1983b)
Skim milk or cream	Natural	Pasteurization at 64°C for 30 min	No obvious change	Wiseman et al. (1983)

that heat treatments similar to those used with dairy foods (e.g. pasteurization and sterilization) do not cause an appreciable change in the amount of AFM_1 in these products.

2.1.2 Storage at low temperature

Studies on stability of AFM_1 in milk during cold storage have given variable results. McKinney *et al.* (1973) stored naturally-contaminated raw milk at $0°C$. The amount of AFM_1 in milk steadily decreased during storage. Losses after 12 days of storage were 89–100% of the initial amount of toxin in milk. A study that was done later by investigators from the same group (McKinney & Cavanagh, 1977) raises doubts about the validity of these findings. Another study (Kiermeier & Mashaley, 1977) indicates that storage of naturally-contaminated milk at $4°C$ for 1–3 days caused some reduction (11–25%) in the amount of AFM_1. Stoloff *et al.* (1975), however, recovered essentially all AFM_1 from artificially-contaminated raw milk that was stored at $4°C$ for up to 17 days.

Studies on the stability of AFM_1 during frozen storage of contaminated milk also gave variable results. McKinney *et al.* (1973) found that the concentration of AFM_1 in naturally-contaminated raw milk decreased steadily during frozen storage at $-18°C$. During 120 days of frozen storage *c.* 87% of AFM_1 in a milk sample was lost. McKinney & Cavanagh (1977), however, found that recovery of AFM_1 from frozen contaminated raw milk is greatly affected by the method of extracting the toxin from this product. The amount of AFM_1 recovered from contaminated raw milk that was stored at $-10°C$ for 25 months was down to 20% when they used an AOAC extraction procedure, but they recovered all of the toxin when a different extraction procedure was followed.

According to results of other investigators (Kiermeier & Mashaley, 1977), freezing artificially-contaminated milk at $-18°C$ and storing it for 6 days at that temperature decreased its content of AFM_1 by 32%. When they used naturally-contaminated milk, the comparable loss was 2–8%. In contrast to these results, Stoloff *et al.* (1975) found that the concentration of AFM_1 in naturally-contaminated raw milk stored at $-18°C$ did not decrease during the first 68 days of frozen storage. Loss in recovery of AFM_1 during frozen storage for 120 days was 45% of the original amount. McKinney *et al.* (1973) observed that AFM_1 in curd and whey was not affected by frozen storage for four months. Wiseman & Marth (1983*d*) found AFM_1 to be stable during frozen storage of contaminated ice-cream and sherbet for eight months.

Because of their variability, a completely firm conclusion can not be made from results of these studies. However, storage of frozen contaminated milk and other dairy products for a few months does not appear to affect their content of AFM_1.

2.1.3 Manufacture of cultured dairy products

Cultured dairy products are manufactured by heat treating the milk and adding a starter culture to initiate the fermentation. Acid develops during the fermentation, then the fermentation is terminated by cooling the product to the desired storage temperature. Wiseman & Marth (1983*b*) found that heat treatments, similar to those used in making yogurt or kefir, did not decrease the concentration of AFM_1 in milk. To study the effect of an acidic environment similar to that encountered in cultured milks, Wiseman & Marth (1983*b*) incubated AFM_1-contaminated phosphate–citrate (McIlvaine's) buffer solutions of pH 4–6·6 at 28°C for up to 96 h. This acid treatment did not appreciably affect the concentration of AFM_1 in the buffer solutions. These results are consistent with those of an earlier study on stability of AFB_1 in an acidic environment. Doyle & Marth (1978*a*) found that AFB_1 was stable at pH 4. At lower pH values, however, degradation of AFB_1 was noted.

Van Egmond *et al.* (1977) made yogurt from naturally-contaminated milk. They observed a slight increase in AFM_1 concentration (*c.* 9%) during yogurt-making, but the AFM_1 content of yogurt did not change during seven days of storage at 7°C. Wiseman & Marth (1983*a*) used naturally-contaminated milk to manufacture yogurt, cultured buttermilk and kefir. A large variability in results of this study makes it difficult to assess the behavior of the toxin in these products. However, the concentration of AFM_1 in yogurt after 6 weeks, and in some batches of cultured buttermilk after two weeks of storage at 7°C was approximately the same as that in the initial milk. Concentration of AFM_1 in kefir after two weeks of storage at 7°C was *c.* 40% lower than that in milk from which it was made.

2.2 Processes that Separate Milk Components

2.2.1 Concentration and drying of milk

Partial or complete removal of water from milk results in such products as evaporated, concentrated, or dried milks. Although removal of water is usually done by heat, some products in this category are processed without heating (e.g. freeze-dried milk). When studying the stability of AFM_1 in these products, it is important to account for the increase in concentration of the toxin in the final product resulting from removal of water.

Researchers usually assumed that freeze-drying of milk does not alter its content of AFM_1 (e.g. Purchase & Steyn, 1967; Purchase *et al.*, 1972). Wiseman & Marth (1983c) reported no loss in amount of AFM_1 caused by freeze-drying naturally-contaminated skim milk or buttermilk, or by storage of these freeze-dried products for four months at 22°C. Kiermeier & Mashaley (1977), however, found that freeze-drying naturally-contaminated milk followed by storing the product for 15 days at 18°C caused a loss of 9·6% of the amount of AFM_1 that was originally present in milk.

Earlier in this discussion, it was mentioned that heating milk at temperatures normally used to process dairy products does not affect the amount of AFM_1 in the milk. During concentration or drying of milk, however, factors other than heat may affect the stability of AFM_1. Removal of water from milk serves to concentrate milk solids and such contaminants as AFM_1. This may make the toxin more susceptible to effects of oxygen, light or other destabilizing factors. However, this assumption was not conclusively proven by experimental results.

Purchase *et al.* (1972) reported losses in amount of AFM_1 caused by evaporating and drying of milk that ranged from 61% to 86%. Other researchers reported much smaller losses of AFM_1 caused by such processes. Kiermeier & Mashaley (1977) found that concentrating naturally-contaminated milk did not affect its content of AFM_1, but spray-drying decreased the amount of AFM_1 by 12·6%. Allcroft & Carnaghan (1962, 1963) found that roller-drying caused no change in toxicity of AFM_1 as determined by duckling bio-assay. Patel *et al.* (1981) used naturally-contaminated buffaloes' milk to produce Khoa. Khoa is a popular concentrated milk product in India. It is made by heating milk in a shallow pan over a steady flame until the product is viscous (Lembhe *et al.*, 1981). In the study of Patel *et al.* (1981), AFM_1-contaminated milk was heated for 3–4 h before Khoa was made. According to their results, there was no evidence for loss of AFM_1 during making this product.

2.2.2 Manufacture of cream and butter

Manufacture of cream involves separation of some of the aqueous phase of milk thereby concentrating the fat globules in the remainder of the liquid. In making butter from cream, fat globules fuse during the churning process and serum is again separated. AFM_1 is mainly soluble in the aqueous phase of milk or possibly adsorbed to casein particles (see the discussion on this subject in Section 2.2.4). This suggests that cream should contain a lower concentration of AFM_1 than milk from which it is made, and butter should contain even lower concentration of toxin than the cream from which it is made.

Results of most of the few studies that dealt with partition of AFM_1 during cream- and butter-making support the situation just described. Grant & Carlson (1971) mixed equal volumes of skim milk and cream that contained measured amounts of AFM_1. The mixture was allowed to equilibrate, and then was separated into equal volumes of cream and skim milk. The researchers recovered 75% of the added toxin from the skim milk fraction. They concluded that during separation of aflatoxin-contaminated milk, the cream fraction would contain *c.* 10% of the toxin initially present in milk. The same investigators added AFM_1 to cream that was used to make butter. The amount of AFM_1 that was recovered from the butter was only *c.* 10% of that of the added toxin. Kiermeier & Mashaley (1977) produced cream and butter from naturally-contaminated milk. Their results indicate that concentration of AFM_1 in cream was 32% lower than it was in the milk from which cream was made. Butter contained 18% and buttermilk 81% of the AFM_1 in cream from which the butter was made. The concentration of AFM_1 in butter was 58% lower than it was in the cream from which the butter was made, and the ratio of concentration of AFM_1 in butter to that in buttermilk was 0·3. When Stubblefield & Shannon (1974) made butter from artificially-contaminated cream, 16% of AFM_1 in cream was recovered from butter and 86% from buttermilk. Concentration of AFM_1 in butter was *c.* 58% lower than it was in the cream from which butter was made. Van Egmond & Paulsch (1986) processed naturally-contaminated milk into cream (20 or 40% fat) and butter. Concentration of AFM_1 in cream was 33–40% lower than it was in milk, and toxin concentration in butter was 56–61% lower than it was in cream. Wiseman *et al.* (1983), however, found that distribution of AFM_1 during production of cream from naturally-contaminated milk was similar to the partitioning by weight, of whole milk into cream and skim milk.

It is evident from the data just described that a small proportion of AFM_1 in milk is carried-over to cream, and yet a smaller proportion to butter. The remainder of the AFM_1 in milk, however, remains in skim milk and buttermilk.

2.3.3 Manufacture of cheese
Conversion of milk into cheese involves some basic steps that apply to most cheese-making operations. These include precipitation of milk

proteins (mainly casein), removal of whey and ripening or cold storage of newly-made cheese. Precipitation of milk proteins is usually accomplished by acids, enzymes, heat, or certain combinations of these treatments. This first step initiates separation of whey (syneresis). Removing most of the whey is enhanced by the physical, chemical and microbial changes that occur in the curd during cheese-making. Curd may be pressed mechanically to remove more whey. Then cheese may be stored for a few days and eaten fresh or ripened for a few weeks to several years. While only a few physical and chemical changes occur in the curd after making soft cheeses to be consumed fresh, extensive changes may occur in ripened cheeses. When milk contaminated with AFM_1 is made into cheese, the fate of the toxin is likely to depend on how these major manufacturing steps are applied.

Shortly after discovery of aflatoxins, Allcroft & Carnaghan (1962, 1963) fractionated a naturally-contaminated milk into rennet-pre-cipitated casein, whey proteins (albumins and globulins) and the remaining liquid. In their assay for toxin (probably AFM_1), only the rennet-precipiated fraction was toxic to ducklings and its toxicity was similar to that of the whole milk. In contrast to these findings, Purchase *et al.* (1972) found that cottage cheese made from naturally contaminated milk contained no AFM_1, but the AFM_1 was present in the whey. Results of these early studies represent two extremes that were not confirmed by later studies. Since then several studies have indicated that when cheese was made from AFM_1-contaminated milk, the toxin was detected in both curd and whey. Results of these more recent studies may enable us to explain the findings of Allcroft & Carnaghan; a higher concentration of AFM_1 was in the curd than the whey and the concentration of AFM_1 in the whey was smaller than detection limits of the duckling assay procedure they followed. Results of Purchase *et al.* (1972), however, contradict those of all the other studies that we encountered, and thus were not confirmed.

Studies on the fate of AFM_1 during cheese-making attempted to answer the following questions. (a) Does AFM_1 degrade during cheese-making? (b) How much AFM_1 is carried-over to cheese from milk? (c) Is AFM_1 partitioned disproportionately between the whey and the curd? In the following discussion, we will review results of various studies on this subject to seek answers to the questions.

2.2.3.1 Stability of aflatoxin M_1 during cheese-making. In this discussion, the cheese-making procedure will be divided into two major phases; (a) conversion of milk into pressed curd, and (b) ripening or storage of this newly-made cheese. Stubblefield & Shannon (1974) made several varieties of cheese from artificially-contaminated milk. They found that 88–111% of AFM_1 added to milk was recovered in cheese-curd and liquid fractions. Kiermeier & Buchner (1977a) used naturally-contaminated milk to make three different types of cheese. Curd and whey contained 94–99% of AFM_1 in milk. Other investigators obtained similar results (McKinney *et al.*, 1973; Van Egmond *et al.*, 1977). These studies show that AFM_1 is not degraded during the first phase of cheese-making.

Results of studies on the fate of AFM_1 in cheese during ripening are variable. The concentration of AFM_1 in Camembert and Tilsit cheeses increased during the early stage of their ripening, and then decreased to approach the concentration of the toxin in cheese at the beginning of ripening (Kiermeier & Buchner, 1977b). Similar observations were reported in studies on Cheddar (Brackett & Marth, 1982b) and brick (Brackett *et al.*, 1982) cheeses. An inadequate method of analysis for AFM_1 was thought to cause the unexpected trend (Brackett & Marth, 1982b; Brackett *et al.*, 1982). AFM_1 in Gouda (Van Egmond *et al.*, 1977) or Mozzarella (Brackett & Marth, 1982c) cheeses did not change appreciably in amount during ripening of these cheeses for six or four months, respectively. Concentration of AFM_1 in Parmesan cheese decreased during the first few months of ripening, then slowly increased (Brackett & Marth, 1982c). In spite of these discrepancies, it can be concluded that AFM_1 does not seem to be degraded during ripening of most cheeses.

2.2.3.2 Carry-over of aflatoxin M₁ into cheese. The general consensus that one can draw from the numerous studies on the fate of AFM_1 during cheese-making is that cheese contains less AFM_1 than the milk from which it is made (Table 2). Most, if not all, of this loss results from the partition of AFM_1 between curd and whey. Results in Table 2 show that the carry-over of AFM_1 from milk to cheese ranged between 0–118%. Most investigators, however, reported the amount of AFM_1 in cheese was in the range of 40–60% of the amount in milk. Most of the remaining AFM_1 was in the whey.

2.2.4 Association of aflatoxin M₁ with casein

When cheeses are made from AFM_1-contaminated milk, toxin is partitioned between curd and whey. If that partition is explained on the basis of solubility of aflatoxin in water, a higher concentration of the toxin in whey than in curd is expected. Most research results, however,

TABLE 2
Partition of aflatoxin M_1 during cheese-making

Cheese type	Source of AFM_1 in milk	Carry-over[a] (%)	EF^b	Reference
1. Soft cheeses				
Experimental rennet curd	Natural	c. 42	c. 2·1	McKinney et al. (1973)
Experimental:	Artificial			Kiermeier & Buchner (1977b)
Rennet curd			2·5–2·8	
Acid curd			2·7	
Unripened	Natural	59	3·0	Kiermeier & Buchner (1977a)
Unripened	Natural	59	3·2	Kiermeier & Buchner (1977b)
Cottage	Artificial	c. 50	c. 2·5	Grant & Carlson (1971)
Cottage	Natural	0	0	Purchase et al. (1972)
Cottage:	Artificial			Stubblefield & Shannon (1974)
Short-set		40	2·7	
Long-set		45	3·0	
Cottage	Natural	12	0·83	Stoloff et al. (1975)
Cottage	Natural	118	8·1	Applebaum & Marth (1982a)
Ricotta	Artificial	30	1·0	Stubblefield & Shannon (1974)
Queso Blanco	Artificial	50	4·3	Stubblefield & Shannon (1974)

Camembert	Natural	58	2·9	Kiermeier & Buchner (1977a)
Camembert	Natural	57	3·3	Kiermeier & Buchner (1977b)
Mozzarella	Natural	73	7·1	Brackett & Marth (1982c)
2. Semi-hard cheeses				
Tilsit	Natural	51	2·6	Kiermeier & Buchner (1977a)
Tilsit	Natural	47	3·7	Kiermeier & Buchner (1977b)
Brick	Natural	12	1·7	Brackett et al. (1982)
3. Hard cheeses				
Gouda	Natural	47	3·9	Van Egmond et al. (1977)
Colby	Artificial	56	5·6	Stubblefield & Shannon (1974)
Swiss	Artificial	44	4·4	Stubblefield & Shannon (1974)
Cheddar	Artificial	47	4·7	Stubblefield & Shannon (1974)
Cheddar	Natural	45	4·3	Brackett & Marth (1982b)
Parmesan	Natural	62	5·8	Brackett & Marth (1982c)

[a] $\% \text{ Carry-over} = \dfrac{\text{Amount of AFM}_1 \text{ in cheese-curd}}{\text{Amount of AFM}_1 \text{ in cheese-milk}} \times 100$

[b] $\text{Enrichment factor} = \dfrac{\text{Concentration of AFM}_1 \text{ in cheese-curd}}{\text{Concentration of AFM}_1 \text{ in cheese-milk}}$

indicate a higher concentration of AFM_1 in curd than in whey. Aflatoxins are not fat-soluble; accordingly, inclusion of fat in the curd (during clotting of milk) can result only in reducing the concentration of the toxin in the curd. Only a few investigators addressed the question of possible association of AFM_1 with whey proteins (Allcroft & Carnaghan, 1962, 1963; Blanc *et al.*, 1983). Results of their studies gave no evidence for that association. Association of AFM_1 with casein seems to be the only plausible explanation for the findings discussed earlier. Brackett & Marth (1982*a*) conducted a study to prove this assumption. They separated casein from milk, then dialyzed it against simulated milk ultrafiltrate (SMUF) that contained AFM_1. After 24 h of dialysis at $7°C$, the casein suspension contained 2·5- to 2·9-fold more toxin than in the SMUF. The study of Blanc *et al.* (1983) also provided evidence for association of AFM_1 with casein. They separated casein from artificially-contaminated skim milk by acid precipitation, rennet coagulation, and centrifugation. They found that the concentration of AFM_1 in separated casein was 2·5-, 3·6-, and 4·6-fold more than it was in skim milk, respectively.

Association of AFM_1 with casein causes the cheese-curd to contain a higher concentration of the toxin than the whey. We calculated or estimated the ratio of concentration of AFM_1 in curd and whey using results from a few studies. This ratio was 6·2–6·4 in an uncured, 5·8–6·3 in Camembert, 4·9–7·2 in Tilsit (Kiermeier & Buchner, 1977*a, b*), 6·5 in Gouda (Van Egmond *et al.*, 1977), and 9·0 in Cheddar cheese (Stubblefield & Shannon, 1974). The association of AFM_1 with casein is also manifested in a higher concentration of AFM_1 in cheese than in the milk from which the cheese is made. This relationship can be expressed as an enrichment factor (EF) for AFM_1 during cheese-making. We used results of earlier studies to calculate the EF value of AFM_1 during cheese-making, unless the value of this parameter was included in the study (Table 2). Values of cheese yield that we used in our calculations were those reported by Kosikowski (1977) and other literature, unless this information was included in the publication. Data on EF (Table 2) show that the concentration of AFM_1 is 2·5- to 3·3-fold higher in many soft cheeses and 3·9- to 5·8-fold higher in hard cheeses than that in milk from which these cheeses were made.

Currently available research results are sufficient to allow some speculation about the nature of the association between milk proteins and AFM_1 (Brackett & Marth, 1982*b*; Van Egmond, 1983). Most of the AFM_1 in milk can be recovered from cheese and whey (see Section

2.2.3.1). AFM₁ associated with casein can be recovered by thorough washing with water (Blanc *et al.*, 1983). This suggests the association of the toxin with casein is not chemical in nature. AFM₁ favors casein more than the separated aqueous phase (whey), but can be removed from this protein when extracted with organic solvents. These observations may be explained by a hydrophobic interaction between casein and AFM₁. The casein molecule contains hydrophobic regions on its surface (Dosako *et al.*, 1980). These regions may provide adsorption sites for the toxin.

Literature that directly addresses the question of association of AFM₁ with casein is limited. However, we believe that some of the behavior of AFM₁ in milk and dairy products can be interpreted by that association. Some of the factors that we believe affect association of AFM₁ with casein are considered in the following paragraphs.

2.2.4.1 Heat treatment. Brackett & Marth (1982*b*) noted that heating AFM₁-contaminated Cheddar cheese for 20 min at 90°C increased recovery of the toxin by 34%. Processed cheese made from that contaminated cheese contained 31–67% more toxin than was expected. The same researchers (Brackett & Marth, 1982*c*) made Mozzarella cheese from naturally-contaminated milk. The EF for AFM₁ in this cheese was 7·1, a value higher than reported in most studies (Table 2). During the manufacture of Mozzarella cheese, curd is heated in water at 80°C to give this cheese its characteristic 'rubbery' texture. This heat treatment may have affected the association of AFM₁ with casein, causing an apparent increase in the amount of toxin recovered from this cheese. Applebaum & Marth (1982*c*) found that more AFM₁ was degraded by H_2O_2 (6% in milk) in pasteurized than in raw AFM₁-contaminated milk. Adsorption of AFM₁ to casein in raw milk may have made the toxin less accessible to the degrading agent.

Stubblefield & Shannon (1974) made Ricotta cheese from artificially contaminated milk. To manufacture this cheese, milk proteins were precipitated with heat (80°C) at pH 5·9–6·0, and then cooked at *c.* 100°C. The estimated EF in this cheese is 1 (Table 2). They manufactured other cheeses that received lower heat-treatments, but gave a much higher EF value than that of Ricotta. This suggests no binding of toxin to casein in this cheese. It is likely that heat affected the casein molecules in a way that disturbed the hydrophobic regions on the casein surface, accordingly decreased adsorption of toxin by those regions.

2.2.4.2 Cheese ripening and proteolysis. Proteolysis of casein
increased the recovery of AFM_1 from naturally-contaminated milk by
31% (Brackett & Marth, 1982*a*). It is likely that the apparent increase in
concentration of AFM_1 in Cheddar cheese during ripening (Brackett &
Marth, 1982*b*) resulted from the partial proteolysis of casein. In contrast,
Mozzarella cheese which undergoes minimal ripening during storage
(Reinbold, 1963), contained a nearly constant concentration of AFM_1
during four months of storage (Brackett & Marth, 1982*c*). Proteolysis of
casein may affect hydrophobic regions on casein molecules, thus
releasing AFM_1 associated with casein molecules.

2.3 Possible Sources of Variability in Results

It is difficult to make firm conclusions about the fate of AFM_1 during
processing of dairy products. Data from different sources are variable
and sometimes conflicting. Factors that may have contributed to this
large variability in results, and thus to data that fail to agree with each
other are discussed in the following paragraphs.

2.3.1 Inadequate methods of analysis

Problems are commonly encountered when measuring concentrations
of AFM_1 in dairy products, and some of the variation in results can be
attributed to the method of analysis employed (Stoloff, 1980; Wiseman
& Marth, 1983*a*). When AFM_1 is in milk, it is usually present at $\mu g/kg$
levels. Detecting such small amounts of toxin requires a reliable method
of analysis and experienced analysts. The main steps in analyzing dairy
products for AFM_1 are extraction, purification, measurement and
confirmation. Extraction of AFM_1 from milk and dairy products
involves mixing (an) organic solvent(s) with the product. Solvents ideal
for solubilization of AFM_1 tend to form emulsions with liquid milk
(Stubblefield & Shannon, 1974). In the past, methods did not include
enough measures to prevent formation of such emulsions. Therefore,
unknown amounts of toxin could have been lost in the emulsion.
Extracts are usually purified using liquid chromatography, and further
loss of AFM_1 may occur. Methods of measuring the amount of toxin in
the studies presented in this chapter vary from duckling bio-assay for
toxicity to the sophisticated use of liquid chromatography (HPLC). The
sensitivity of these methods varies greatly. Some research workers did

not confirm the identity of the toxin they measured while others used various confirmatory tests.

Some results of studies on fate of AFM_1 during cheese ripening seem to represent changes in the recovery of toxin by the method during the different phases of the study rather than real changes in the level of AFM_1 in cheese (Brackett & Marth, 1982*b*). A method developed to measure aflatoxin M_1 in a certain dairy product may give inadequate recovery of the toxin when used with other products. There are some interfering fluorescent materials that occasionally appear in milk and dairy products. These materials may co-chromatograph with aflatoxin M_1 when certain analytical methods are used (Miller *et al.*, 1980; Price *et al.*, 1981; Wiseman & Marth, 1983*d*). McKinney & Cavanagh (1977) made cheese from AFM_1-contaminated milk, and followed partition of the toxin into curd and whey using two different methods of analysis. They found that 75% of AFM_1 occurred in the curd and 25% in the whey when one extraction method was used. When a different method was employed, they found 40% of the toxin in curd and 60% in whey.

2.3.2 Kind of contamination

Some researchers (Kiermeier & Mashaley, 1977; Applebaum & Marth, 1982*b*) found that the kind of contamination (natural vs artificial) causes variability in data on stability of AFM_1 in some dairy products. It is likely that toxin is more homogeneously distributed in naturally-contaminated than in artificially-contaminated milk. The way milk is spiked with toxin usually involves mixing the toxin (present in a small volume of an organic solvent) with milk. Organic solvents best suited for dissolving aflatoxins (e.g. chloroform) are immiscible with aqueous media like milk. If the aflatoxin solution forms an emulsion in milk, only thorough shaking of the mixture may eventually cause most of aflatoxin to migrate to the large volume of the surrounding aqueous phase. Therefore, irregularity in results of studies on milk that has different kinds of contamination is conceivable.

2.3.3 Other factors

Factors other than those just mentioned may play a role in the stability of AFM_1 in dairy products. Type of feed given to lactating animals and seasonal variation in milk composition may contribute to instability of the toxin in milk during processing. The oxidation–reduction potential of milk and exposure of milk to light are examples of other factors that may contribute to instability of AFM_1 in milk.

3 DEGRADING AFLATOXIN M₁ IN MILK BY SPECIAL TREATMENTS

From the discussion thus far it is clear that AFM_1 is fairly stable during processing of toxin-contaminated milk into various dairy products. Normal treatments during processing of milk, such as use of heat, and development of acid during fermentation do not destabilize the toxin. Removal of whey during cheese-making eliminates an appreciable amount of the AFM_1 in milk, but the cheese produced usually contains a higher concentration of the toxin than did the milk from which it was made. Use of special treatments designed to free milk from the toxin seems to be a reasonable approach to solve the problem. Treatments that may physically remove or chemically degrade or destabilize AFM_1 in milk have been investigated by only a few researchers. Results of these studies only provide preliminary answers to a question that needs much broader and more intensive investigation.

3.1 Chemical Treatments

3.1.1 Sulphites and bisulphites
Sulphites and bisulphites are multi-purpose food additives that are used with a variety of foods (Joslyn & Braverman, 1954; Roberts & McWeeny, 1972). Several studies indicate that sulphites or bisulphites can degrade AFB_1 in buffer solutions (Doyle & Marth, 1978*b*, *c*) or in corn (Moerck *et al.*, 1980; Hagler *et al.*, 1982). On the basis of these observations, Applebaum & Marth (1982*c*) investigated the possible use of sulphites to degrade AFM_1 in milk. Raw milk that was naturally contaminated with AFM_1 was treated with 0·4% potassium bisulphite at 25° C for 5 h. The concentration of AFM_1 in milk decreased by 45% from this treatment. A higher concentration of bisulphite was less effective in eliminating AFM_1 from milk.

3.1.2 Hydrogen peroxide
Currently, food regulations in some countries permit adding hydrogen peroxide (H_2O_2) to milk as a preservative. In the USA, 0·05% of H_2O_2 may be added to milk used to manufacture certain cheeses (Anonymous, 1977). Researchers (Sreenivasamurthy *et al.*, 1967) used H_2O_2 to successfully detoxify aflatoxins in defatted peanut meal. Best results

were obtained when the pH of the suspension (containing H_2O_2) was adjusted at 9·5, heated at 80° C for 30 min, cooled, and then acidified.

Although the concentration of H_2O_2 and the heat treatment employed in the study just mentioned are higher than would be acceptable for milk, a study was conducted to explore the possibility of degrading AFM_1 in milk by H_2O_2 (Applebaum & Marth, 1982*b*). In this study, AFM_1 (0·7–1·7 μg/kg) in naturally-contaminated milk was not affected by the presence of 1% H_2O_2 at 30° C for 30 min. Detectable losses of AFM_1 were achieved by using a higher concentration of H_2O_2 and/or a higher temperature during treating of the milk (Table 3).

In an attempt to reduce the amount of H_2O_2 needed to degrade AFM_1 in milk, combinations of H_2O_2 and other additives with or without heat treatments were studied (Applebaum & Marth, 1982*b*). Combinations of H_2O_2 and riboflavin were more effective in degrading AFM_1 in milk than were the individual additives. Adding lactoperoxidase to milk did not affect its content of AFM_1. However, use of H_2O_2 plus lactoperoxidase made it possible to use less H_2O_2 (0·05 and 0·1%) than was needed with other combinations of H_2O_2 and additives and still degrade 47–52% of AFM_1 in milk.

TABLE 3
Degradation of AFM_1 (0·7–1·7 μg/kg) in naturally-contaminated milk by hydrogen peroxide (H_2O_2) with or without added riboflavin or lactoperoxidase (Applebaum & Marth, 1982*b*)

Treatment	Time (min)	Temperature (°C)	AFM_1 degraded (%)
1% H_2O_2	30	30	0
1% H_2O_2	30	63	11
6% H_2O_2	30	30	43
6% H_2O_2	30	63	71
6% H_2O_2 + 3·2 mM riboflavin	15	30	100
1% H_2O_2 + 0·5 mM riboflavin + heating at 63°C for 30 min	30	30	98
0·05% H_2O_2 + 5 UL/ml[a]	140	23	47
0·1% H_2O_2 + 5 UL/ml	140	23	52

[a]Unit of activity of lactoperoxidase per ml of the reaction mixture (see Applebaum & Marth, 1982*b* for details).

3.2 Physical Treatments

3.2.1 Adsorption on particulate materials

Hypothetically, removing AFM_1 from milk by adsorption onto particulate material has some advantages over other methods for eliminating the toxin from milk. Such a procedure does not usually involve degrading the toxin; accordingly, milk free from toxin-degradation products and safe for consumption may be produced. The added material may be easily separated from milk after the substance adsorbs the toxin. The particulate material, however, may adsorb some valuable components (e.g. proteins and vitamins) from milk, thus reducing its nutritive value. Economic and regulatory consequences of such a procedure should be assessed carefully before it can be used in the dairy industry.

A study on adsorption of AFB_1 indicated that vermiculite, bentonite and other clays effectively removed the toxin from buffer solutions (Masimango *et al.*, 1978). Other investigators (Decker & Corby, 1980) found that activated charcoal was highly effective in removing AFB_1 from a complex medium. When activated charcoal was included in AFB_1-contaminated feed, it reduced chronic toxicity of AFB_1 to chickens (Dalvi & Ademoyero, 1984). The structural similarity between AFB_1 and AFM_1 prompted Applebaum & Marth (Applebaum & Marth, 1982*c*) to explore possible adsorption of AFM_1 in milk onto bentonite. Bentonite (0·5–2%) was added to naturally-contaminated milk that contained 3–6 μg of AFM_1/kg. Bentonite (5% in milk) adsorbed 89% of AFM_1 in 60 min at 25°C. According to Charm (personal communication), a charcoal column designed to remove antibiotics from milk also was capable of removing AFM_1.

3.2.2 Treatment with ultraviolet energy

Ultraviolet (UV) radiation was used in the past to enrich milk with vitamin D. The process was replaced with the less expensive direct addition of the vitamin to milk. Some aflatoxins can be degraded by long-wave UV radiation (Andrellos *et al.*, 1967; Aibara & Yamagishi, 1970). In view of these facts, studies (Yousef & Marth, 1985*a*; 1986) were done to investigate the possibility of using UV energy to degrade AFM_1 in milk.

In exploratory experiments (Yousef & Marth, 1985*a*), raw whole milk was spiked to contain 0·5 or 1 μg of AFM_1/kg. UV sources, with different power ratings (60 or 100 watts), that emit long-wave UV radiation were used. Stationary layers of milk or milk circulating in a glass reaction vessel were exposed to radiation from the UV lamps. In all instances, exposure of contaminated milk to UV radiation caused AFM_1 to degrade. In some experiments, degradation of the toxin to below the detection limit (0·01 μg/kg) of the method used for analysis (Yousef & Marth, 1985*b*) was achieved. The magnitude of degradation (3·6–100%) depended on the length of time (2–60 min) the milk was exposed to UV radiation, design of the experiment, volume of treated milk, and presence of H_2O_2 in milk.

To develop a practical process that could be used by the dairy industry to degrade AFM_1 in milk, knowledge is needed about various factors that affect degradation of AFM_1. A study (Yousef & Marth, 1986) was conducted to develop such knowledge. In that study, raw whole milk was spiked with AFM_1 to contain 1 μg of the toxin/kg. A thin layer of this milk (1-mm deep) at a controlled temperature was subjected to a low-energy (60 watts) UV source. The main wavelength emitted by the UV lamp was at 365 nm. Results indicated that AFM_1 was degraded during UV radiation of milk in a pattern similar to that of first-order reaction kinetics. Raw whole AFM_1-contaminated milk was heated at 90° C for 10 min to inactivate most enzymes occurring naturally in milk. After cooling, this milk was treated with UV energy. UV radiation degraded AFM_1 in preheated milk to about the same degree as in the raw milk. Thus enzymes in milk do not seem to participate in the process of degrading AFM_1 by UV radiation.

3.2.2.1 Temperature.

3.2.2.1 Temperature. Results of the study just described (Yousef & Marth, 1986) indicated that the temperature of milk during exposure to UV radiation affected the extent of AFM_1 degradation. Higher temperatures (in the range of 5–65° C) were more favorable for the degradation reaction than were lower ones. Degradation of AFM_1 was enhanced more by increasing the temperature of milk from 5 to 25° C than from 25 to 65° C (Fig. 1).

Further studies on kinetics of inactivation of AFM_1 by UV radiation were done using aqueous solutions rather than milk (Yousef & Marth, 1987). Thin layers (1 mm deep) of aqueous solutions of AFM_1 were subjected to UV radiation at different temperatures. Reaction rate constants for conversion of AFM_1 to a suggested reaction product were calculated for treatments at 0, 30 and 60° C. Data show that the activation energy for this reaction was 2·3 kcal/mol, a small value compared with that of most chemical reactions. The degradation

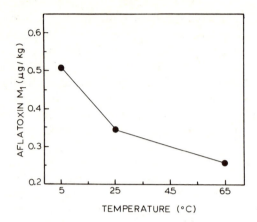

Fig. 1. Degradation of AFM_1 in milk held at different temperatures and treated with UV radiation for 20 min (Yousef & Marth, 1986). Initial concentration of AFM_1 in milk was $0.84\,\mu g/kg$. UV source was a 60-watt lamp with a main wavelength at 365 nm. The lamp was 3.9 cm above a thin layer (c. 1 mm deep) of milk. Data are average values of three trials.

reaction had $Q_{10} = 1.13$ (Q_{10} is the ratio of reaction rate constant at a given temperature, $t\,°C$, to that at $(t - 10)°C$). Most chemical reactions have a Q_{10} value of 2–3. It is likely that the AFM_1 molecule derives its activation energy from UV radiation rather than the heat energy of the solution. Although increasing the temperature of the medium in which AFM_1 was treated with UV radiation increased the extent of degradation, this increase was much less than that usually experienced in non-photochemical reactions.

3.2.2.2 pH. Altering the pH of milk causes physical changes that make it impractical to study the effect of this parameter on degradation of AFM_1 by UV radiation. Therefore, aqueous solutions of AFM_1 were used for that purpose (Yousef & Marth, 1987). AFM_1 was dissolved in citrate buffer (0.01 M) that had a pH value of 3, 5 or 7, and solutions were subjected to UV radiation. Data show that the rate of degradation of the toxin was not affected by pH of the reaction mixture.

3.2.2.3 Presence of oxidizing agents. Many countries, including the USA, permit limited use of H_2O_2 as a food additive. Consequently, Yousef & Marth (1985a, 1986) investigated degradation of AFM_1 by UV radiation when milk contained H_2O_2. Four hundred grams of AFM_1-

contaminated milk (0·5 μg of AFM_1/kg) was circulated in a glass reaction vessel and treated with UV radiation (100-watt UV source) for 10 min (Yousef & Marth, 1985*a*). This treatment decreased the concentration of AFM_1 by 56%. When the same experiment was repeated using milk that contained 1% H_2O_2, no AFM_1 was detected in the treated milk. AFM_1 in milk was not affected by the presence of 1% H_2O_2 without UV irradiation. Another study (Yousef & Marth, 1986) was done using a different apparatus and experimental design. Results of this study are in Table 4. Presence of H_2O_2 (0·05%) significantly increased the effectiveness of UV radiation for degrading AFM_1 in milk.

Benzoyl peroxide is used in limited applications in the food industry. In the USA, milk used to manufacture blue cheese may be bleached with benzoyl peroxide (up to 0·02% by weight) (Anonymous, 1977). In spite of its oxidizing power, presence of benzoyl peroxide (0·02% by weight) did not affect AFM_1 in milk. When milk containing the same concentration of the chemical was treated with UV energy, AFM_1 degraded to about the same extent as when the peroxide was absent (Table 4).

3.3 Chemistry of Degradation of Aflatoxin M₁

From the foregone discussion, it is evident that only a few studies were devoted to find methods to degrade AFM_1 in milk. To implement any

TABLE 4
Degradation of aflatoxin M_1 in milk that was treated with additives, ultraviolet (UV) radiation, or both (Yousef & Marth, 1986). A thin layer of raw whole milk (1 mm deep) was exposed for 20 min to a 60-watt UV source that was positioned 3·9 cm above the milk layer

Treatment	AFM₁ recovered (μg/kg)	Degradation (%)
Hydrogen peroxide		
Untreated, contaminated milk	0·835	
UV only	0·328	60·7
Peroxide (0·05%) + UV	0·0912	89·1
Benzoyl peroxide		
Untreated, contaminated milk	0·772	
Peroxide (0·02%) only	0·781	−1·17
UV only	0·346	55·2
Peroxide (0·02%) + UV	0·303	46·9

methods for degrading AFM_1 in the dairy industry, products of toxin degradation should be identified and their safety checked. Developing information to understand the chemistry of the degradation reactions in these methods is not an easy task, and requires more than casual investigations. Currently, only limited information is available on this subject. Therefore, the discussion that follows is largely hypothetical in nature. Because of the structural similarity between AFB_1 and AFM_1, results of studies on degrading AFB_1 will be referred to when information about degrading AFM_1 is lacking.

3.3.1 Aflatoxin M_1 and related compounds

Aflatoxins are a group of chemically related compounds that structurally are coumarin derivatives. Aflatoxin B_1 is the most toxic member of the group. AFM_1 differs from AFB_1 in that a hydroxyl group is present on the difuran moiety of its molecule (Fig. 2). AFM_1 has intermediate polarity that makes it soluble in chloroform and similar organic solvents, soluble in methanol and water, but insoluble in hexane.

The difuran moiety is important for both the toxic and carcinogenic properties of these compounds (Carnaghan *et al.*, 1963; Wogan, 1977). Chemical modification of the dihydrofuran ring of AFB_1 usually results in compounds with less toxicity than that of AFB_1 (Carnaghan *et al.*, 1963; Ciegler & Peterson, 1968; Wei & Chu, 1973). The coumarin moiety

Fig. 2. AFM_1 and related compounds.

constitutes the chromophore part of the aflatoxin molecule. Chemical modification of this part usually results in changes in the fluorescence characteristics of the molecule (Parker & Melnick, 1966; Lee *et al.*, 1974).

Possible reactive sites that may be utilized in degrading AFM_1 in food include the terminal dihydrofuran double bond and the lactone linkage on the coumarin moiety. Degradation reactions at these sites were more extensively studied with AFB_1 than with AFM_1 (Fig. 3). Therefore details about these reactions involving AFB_1 will follow.

Alkaline solutions open the lactone ring of AFB_1 to give a nonfluorescent salt of the resulting hydroxy acid (Parker & Melnick, 1966; Lee *et al.*, 1974). The resulting beta-keto acid is readily decarboxylated. Such reactions were observed when AFB_1 was treated with ammonia (Lee *et al.*, 1974) (Fig. 3), and this principle is used in practice to decontaminate feedstuffs that contain AFB_1. Opening the lactone ring of AFB_1 probably makes the resulting hydroxy acid susceptible to oxidation (Sreenivasamurthy *et al.*, 1967; Trager & Stoloff, 1967).

Fig. 3. Reactions that may be utilized for degradation of AFB_1 (and possibly AFM_1) in foods. See Section 3.3 for the details.

Another important degradation reaction of AFB_1 involves treatment with acids. Strong acids catalyze the addition of a water molecule at the terminal dihydrofuran double bond, thus converting AFB_1 into AFB_{2a} (Fig. 3) (Dutton & Heathcote, 1966; Pons *et al.*, 1972; Ashley *et al.*, 1987). This reaction depends greatly on the pH of the reaction medium (Pons *et al.*, 1972). Wei & Chu (1973) provided evidence for addition of water to the terminal double bond of AFB_1 when an aqueous solution of AFB_1 was irradiated with UV energy. Bisulphites can also react with the terminal dihydrofuran double bond of AFB_1 to form a water-soluble product that has UV absorbance and fluorescence characteristics similar to those of AFB_1 (Hagler *et al.*, 1983) (Fig. 3).

3.3.2 Degradation by bisulphites

Earlier in this chapter (Section 3.1.1), use of potassium sulphite to degrade AFM_1 in milk was discussed. Doyle & Marth (1978*b*) monitored degradation of ^{14}C-labelled AFB_1 in a buffer solution containing bisulphite. The bisulphite solution degraded AFB_1 into products that were insoluble in chloroform but soluble in water. Methanol or citric acid retarded degradation of AFB_1, possibly by inhibiting the oxidation of bisulphite. Oxidation of bisulphite by oxygen seems to generate the bisulphite free radical (Schroeter, 1963), which is a highly reactive species. Doyle & Marth (1978*b*) believed the bisulphite free radical reacted with the double bond of the terminal dihydrofuran moiety of AFB_1 to form sulphonic acid addition products. A study by Hagler *et al.* (1983) provided evidence to support that hypothesis. It may be assumed that degradation of AFM_1 in milk by sulphite compounds (Applebaum & Marth, 1982*c*) follows a mechanism similar to that suggested by Doyle & Marth (1978*b*) and supported by the findings of Hagler *et al.* (1983).

3.3.3 Degradation by hydrogen peroxide

AFM_1 or AFB_1 is relatively stable in dilute solutions of H_2O_2 (Sreenivasamurthy *et al.*, 1967; Applebaum & Marth, 1982*b*; Yousef & Marth, 1985*a*). An aqueous solution containing 10 μg of AFB_1 and 0·03 g of H_2O_2 at pH 7·0 was heated at 80° C for 30 min, then cooled and acidified to pH 4·5. The mixture was extracted with chloroform, and all the added toxin was recovered in the chloroform extract (Sreenivasamurthy *et al.*, 1967). Heating a similar mixture but at higher pH values caused a pronounced reduction in amount of toxin recovered; at pH 9·5, the reduction in extractable AFB_1 was 90%.

Heating AFB$_1$ solutions of high pH values caused the toxin to degrade, but presence of H$_2$O$_2$ dramatically increased the extent of the degradation. The authors (Sreenivasamurthy *et al.*, 1967) assumed the instability of AFB$_1$ in solutions of high pH values resulted from opening of the lactone ring of the AFB$_1$ molecule. This made the molecule more susceptible to oxidation by H$_2$O$_2$.

Studies in our laboratory proved that AFM$_1$ in milk was not appreciably affected by the presence of H$_2$O$_2$ (up to 1%), even when this milk was heated at 63°C for 30 min (Applebaum & Marth, 1982*b*; Yousef & Marth, 1985*a*). This stability of AFM$_1$ to H$_2$O$_2$ was greatly reduced when H$_2$O$_2$-containing milk was treated with lactoperoxidase (Applebaum & Marth, 1982*b*), riboflavin (Applebaum & Marth, 1982*b*), or UV irradiation (Yousef & Marth, 1985*a*, 1986). Odajima (1981) studied oxidation of aflatoxin by a myeloperoxidase–H$_2$O$_2$–chloride system. When AFB$_1$ was treated with this system, absorption of UV by the reaction mixture was inhibited at 365 nm. The study showed that hypochlorous acid generated by the peroxidase–H$_2$O$_2$–chloride system degraded AFB$_1$, and ruled out involvement of singlet oxygen in the oxidation. Results of mass spectrometry suggest that one atom of oxygen was incorporated into the AFB$_1$ molecule, with removal of two atoms of hydrogen. Degradation of AFM$_1$ in milk by H$_2$O$_2$ and lactoperoxidase (Applebaum & Marth, 1982*b*) may be explained by a similar mechanism.

Applebaum & Marth (1982*b*) found that presence of riboflavin and H$_2$O$_2$ in contaminated milk decreased the amount of AFM$_1$ that could be recovered more than did H$_2$O$_2$ or riboflavin alone. They suggested involvement of singlet oxygen in the reaction. Riboflavin can be photosensitized by light to form an electronically excited molecule (triplet state). Riboflavin in this state can react with oxygen, generating singlet oxygen. H$_2$O$_2$ may also be involved in formation of singlet oxygen. Singlet oxygen is a highly reactive electrophyl that may react with the double bond in the terminal dihydrofuran moiety of the AFM$_1$ molecule. Results of a study on interactions of light, riboflavin and AFB$_1$ may provide yet another explanation (Joseph-Bravo *et al.*, 1976). Evidence from that study shows that riboflavin, photosensitized with light, complexes with AFB$_1$. An in-vivo study with rats showed that reaction of AFB$_1$ with the photosensitized riboflavin may have protected the toxin from degradation into carcinogenic metabolites, and thus reduced the incidence of aflatoxin-induced cancer. An in-vitro experiment provided evidence for formation of an AFB$_1$–riboflavin complex in a dimethylsulfoxide (DMSO)–water solvent mixture. Formation of this complex was inhibited by the presence of potassium iodide or ascorbic acid. Results of Applebaum & Marth (1982*b*) may be explained on the basis of the mechanism just described.

3.3.4 Degradation by ultraviolet energy

Studies of Yousef & Marth (Yousef & Marth, 1985*a*, 1986) showed that UV irradiation degraded AFM_1 in milk. Products from the degrading of AFM_1 and remaining in milk are unknown. Monitoring of this degradation reaction in aqueous solutions was done by Yousef & Marth (1987). Aqueous solutions of AFM_1 irradiated with UV energy contained a major fluorescent product. This product (symbolized as AFM_x) was more polar and fluorescent than the parent toxin. When measured with a reverse-phase liquid chromatographic system, AFM_x had a retention time similar to that of an AFM_1–trifluoroacetic acid derivative. Conversion of AFM_1 to AFM_x was not affected by the pH of the solution in the range of 3–7. Further degradation of AFM_x during UV irradiation was inhibited at lower rather than at higher pH values in that range.

AFM_1 in chloroform or as a dry film was subjected to UV irradiation. Most AFM_1 was recovered after irradiation and AFM_x was not detected on the chromatogram (Yousef & Marth, 1987). This illustrates the importance of water for conversion of AFM_1 to AFM_x. Wei & Chu (1973), in an earlier study, demonstrated the role of solvents (including water) in the degradation of AFB_1 by UV irradiation. Based on their observations, they suggested that the terminal double bond at the dihydrofuran ring of AFB_1 was the site of photochemically-induced attack by hydrolytic solvents. Assuming the same concept can be applied to AFM_1, hydration of the double bond at the dihydrofuran moiety of AFM_1 can lead to formation of aflatoxin M_{2a}. Efforts to identify AFM_x (Yousef & Marth, 1987) were hampered by the instability of this product when in the dry form and the lack of enough pure compound for further analysis.

Degradation of AFM_1 in aqueous solutions irradiated with UV energy was monitored at 0, 30 and 60° C (Fig. 4). The reaction was assumed to proceed as follows:

$$AFM_1 \xrightarrow{\ k_1\ } AFM_x \xrightarrow{\ k_2\ } \text{further degradation products}$$

Reaction rate constants (k) and activation energy (E_a) for the steps of

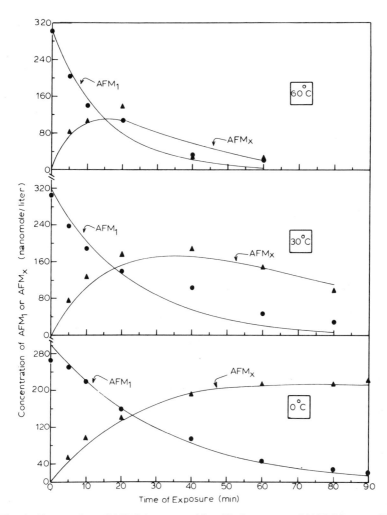

Fig. 4. Conversion of AFM_1 into an unidentified compound (AFM_x) caused by UV irradiation of aqueous solutions of AFM_1 held at different temperatures (Yousef & Marth, 1987).

this reaction were estimated. As indicated earlier, this reaction requires water. Therefore, conversion of AFM_1 to AFM_x in an aqueous solution may be considered to be a pseudo-first-order reaction, and k_1 is the apparent first-order rate constant. The estimated E_a values for the two steps of the reaction are 2·31 and 7·95 kcal/mol, respectively. These E_a values are smaller than those of most non-photochemical reactions. Thus results indicate that degradation of AFM_1 in aqueous solutions is not very sensitive to changes in temperatures, and the second step of the reaction is more sensitive to such changes in temperature than is the first one.

Assuming that the pattern for degrading of AFM_1 by UV irradiation in milk is similar to what was encountered in aqueous solutions (which may not be true), toxicity of AFM_x needs to be assessed. At low temperatures, UV irradiation rapidly degrades AFM_1, but AFM_x accumulates as it is more stable to irradiation than at the higher temperatures. If milk is to be treated with a powerful source of UV irradiation, organoleptic changes may occur in milk (Li & Bradley, 1969). Irradiating milk at high temperatures may further aggravate these changes in quality. Future studies need to determine the optimum conditions for irradiating milk, that maximize elimination of the toxin, with minimum accumulation of reaction products and minimum organoleptic changes in milk.

4 CONCLUSION AND A LOOK TO THE FUTURE

When milk is processed by heat or is fermented, residues of aflatoxin M_1 (AFM_1) in this milk remain stable. Separation of milk components disproportionately partitions the toxin in accordance with its affinity for casein and lack of solubility in fat. Efforts to eliminate, degrade or destabilize AFM_1 in milk are still in the preliminary stage. Only a few ideas have been tested; some of them, however, are promising. Future studies can focus on combining degradation treatments and milk processing. An example for such a venture would be treating milk with UV radiation, followed by ultrafiltration. The lactoperoxidase-thiocyanate system in milk (Reiter & Härnulv, 1984) could be useful for degrading AFM_1. Finally, it may be worthwhile to perfect an enzymatic system to degrade AFM_1, and then clone that system into a useful microorganism such as one of the lactic acid bacteria.

5 SUMMARY

Natural contamination of milk with AFM_1 is occasionally encountered in the dairy industry. There are numerous studies on the fate of AFM_1

during processing of milk, but the results often are quite variable. There is no concrete evidence that cold-storage, freezing, heat-treating, fermenting, concentrating or drying of such a milk appreciably changes its content of AFM_1. Although AFM_1 is not degraded when contaminated milk is manufactured into cheese, cream or butter, it is partitioned into the various products resulting from each of these processes. Cheese made from contaminated milk contains *c.* 40–60%, cream *c.* 10%, and butter less than 2% of the AFM_1 present in milk. When AFM_1-contaminated milk is made into products, resulting cheese contains much higher, but cream and butter contain lower, concentrations of AFM_1 than the milk from which these products are made.

AFM_1-contaminated milk may be treated with physical or chemical procedures that can eliminate the toxin. Some of the treatments studied are those that occasionally are already used in the dairy industry for other purposes and include treating milk with H_2O_2, benzoyl peroxide or UV irradiation. Irradiation with UV energy is a potentially useful technique to degrade AFM_1 in milk. Use of H_2O_2 alone is not effective in degrading the toxin in milk, but it becomes much more effective when milk is also treated with riboflavin, lactoperoxidase or UV irradiation. Some AFM_1 in milk may also be removed by adsorption on certain particulate materials, or degraded by adding bisulphites. Safety and applicability of these techniques have not been tested.

ACKNOWLEDGMENTS

A contribution from the College of Agricultural and Life Sciences, University of Wisconsin-Madison, Madison, WI, USA. Research described in this chapter and done in the Department of Food Science at the University of Wisconsin-Madison was supported by the Dairy Research Foundation, Rosemont, IL, USA, and by Kraft, Inc., Glenview, IL, USA. Preparation of this chapter was supported, in part, by the Wisconsin Milk Marketing Board, Madison, WI, USA.

REFERENCES

Aibara, K. & Yamagishi, S. (1970). Effect of ultraviolet irradiation on the destruction of aflatoxin B₁. In *Proceedings of the First US–Japan Conference on Toxic Microorganisms*, ed. M. Herzberg, US Department of the Interior, Washington, DC, USA, pp. 211–21.

Allcroft, R. & Carnaghan, R. B. A. (1962). Groundnut toxicity: *Aspergillus flavus* toxin (aflatoxin) in animal products: Preliminary communication. *Vet. Rec.,* **74**, 863–4.

Allcroft, R. & Carnaghan, R. B. A. (1963). Groundnut toxicity: an examination for toxin in human food products from animals fed toxic groundnut meal. *Vet. Rec.,* **75**, 259–63.

Andrellos, P. J., Beckwith, A. C. & Eppley, R. W. (1967). Photochemical changes of aflatoxin B_1. *J. Assoc. Off. Anal. Chem.,* **50**, 346–50.

Anonymous (1977). *Code of Federal Regulations, 21: Food and Drugs.* Office of the Federal Register, Washington, DC, USA.

Anonymous (1979). *Aflatoxin and other mycotoxins: An agricultural perspective.* Council for Agricultural Science and Technology, Report No. 80, Ames, IA, USA.

Applebaum, R. S. & Marth, E. H. (1982*a*). Fate of aflatoxin M_1 in cottage cheese. *J. Food Prot.,* **45**, 903–4.

Applebaum, R. S. & Marth, E. H. (1982*b*). Inactivation of aflatoxin M_1 in milk using hydrogen peroxide and hydrogen peroxide plus riboflavin or lactoperoxidase. *J. Food Prot.,* **45**, 557–60.

Applebaum, R. S. & Marth, E. H. (1982*c*). Use of sulfite or bentonite to eliminate aflatoxin M_1 from naturally contaminated raw whole milk. *Z. Lebensm. Unters. Forsch.,* **174**, 303–5.

Ashley, D. L., Orti, D. L. & Hill, R. H., Jr (1987). Proton nuclear magnetic resonance evidence for two configurations of the hemiacetals of aflatoxin B_1 and sterigmatocystin. *J. Agric. Food Chem.,* **35**, 782–5.

Blanc, B., Lauber, E. & Sieber, R. (1983). Binding of aflatoxin to milk proteins. *Microbiologie-Aliments-Nutrition,* **1**, 163–77.

Brackett, R. E. & Marth, E. H. (1982*a*). Association of aflatoxin M_1 with casein. *Z. Lebensm. Unters. Forsch.,* **174**, 439–41.

Brackett, R. E. & Marth, E. H. (1982*b*). Fate of aflatoxin M_1 in Cheddar cheese and in process cheese spread. *J. Food Prot.,* **45**, 549–52.

Brackett, R. E. & Marth, E. H. (1982*c*). Fate of aflatoxin M_1 in Parmesan and Mozzarella cheese. *J. Food Prot.,* **45**, 597–600.

Brackett, R. E., Applebaum, R. S., Wiseman, D. W. & Marth, E. H. (1982). Fate of aflatoxin M_1 in brick and Limburger-like cheese. *J. Food Prot.,* **45**, 553–6.

Carnaghan, R. B. A., Hartley, R. D. & O'Kelly, J. (1963). Toxicity and fluorescence properties of the aflatoxins. *Nature,* **200**, 1101.

Ciegler, A. & Peterson, R. E. (1968). Aflatoxin detoxification: hydroxydihydroaflatoxin B_1. *Appl. Microbiol.,* **16**, 665–6.

Dalvi, R. R. & Ademoyero, A. A. (1984). Toxic effects of aflatoxin B_1 in chickens given feed contaminated with *Aspergillus flavus* and reduction of the toxicity by activated charcoal and some chemical agents. *Avian Dis.,* **28**, 61–9.

Decker, W. J. & Corby, D. G. (1980). Activated charcoal adsorbs aflatoxin B_1. *Vet. Hum. Toxicol.,* **22**, 388–9.

Dosako, S., Kaminogawa, S., Taneya, S. & Yamauchi, K. (1980). Hydrophobic surface areas and net charges of α_{s_1}-, κ-casein and α_{s_1}-casein: κ-casein complex. *J. Dairy Res.,* **47**, 123–9.

Doyle, M. P. & Marth, E. H. (1978*a*). Aflatoxin is degraded at different temperatures and pH values by mycelia of *Aspergillus parasiticus. Eur. J. Appl. Microbiol. Biotechnol.,* **6**, 95–100.

Doyle, M. P. & Marth, E. H. (1978*b*). Bisulfite degrades aflatoxin: effect of citric acid and methanol and possible mechanism of degradation. *J. Food Prot.,* **41**, 891–6.

Doyle, M. P. & Marth, E. H. (1978*c*). Bisulfite degrades aflatoxin: effect of temperature and concentration of bisulfite. *J. Food Prot.,* **41**, 774–80.

Dutton, M. F. & Heathcote, J. G. (1966). Two new hydroxyaflatoxins. *Biochem. J.,* **101**, 21P–22P.

Grant, D. W. & Carlson, F. W. (1971). Partitioning behavior of aflatoxin M in dairy products. *Bull. Environ. Contam. Toxicol.,* **6**, 521–4.

Hagler, W. M., Jr, Hutchins, J. E. & Hamilton, P. B. (1982). Destruction of aflatoxin in corn with sodium bisulfite. *J. Food Prot.,* **45**, 1287–91.

Hagler, W. M., Jr, Hutchins, J. E. & Hamilton, P. B. (1983). Destruction of aflatoxin B_1 with sodium bisulfite: isolation of the major product aflatoxin B_1S. *J. Food Prot.,* **46**, 295–300.

Joseph-Bravo, P. I., Findley, M. & Newberne, P. M. (1976). Some interactions of light, riboflavin, and aflatoxin B_1 *in vivo* and *in vitro*. *J. Toxicol. Environ. Health,* **1**, 353–76.

Joslyn, M. A. & Braverman, J. D. S. (1954). The chemistry and technology of the pretreatment and preservation of fruit and vegetable products with sulfur dioxide and sulfites. *Adv. Food Res.,* **5**, 97–160.

Kiermeier, F. & Buchner, M. (1977*a*). Distribution of aflatoxin M_1 in whey and curd during cheese processing. *Z. Lebensm. Unters. Forsch.,* **164**, 82–6 (in German).

Kiermeier, F. & Buchner, M. (1977*b*). On the aflatoxin M_1 content of cheese during ripening and storage. *Z. Lebensm. Unters. Forsch.,* **164**, 87–91 (in German).

Kiermeier, F. & Mashaley, R. (1977). Influence of raw milk processing on the aflatoxin M_1 content of milk products. *Z. Lebensm. Unters. Forsch.,* **164**, 183–7 (in German).

Kosikowski, F. V. (1977). *Cheese and Fermented Milk Foods*, 2nd Edn. Edwards Brothers, Ann Arbor, MI, USA.

Lee, L. S., Stanley, J. B., Cucullu, A. F., Pons, W. A., Jr & Goldblatt, L. A. (1974). Ammoniation of aflatoxin B_1: isolation and identification of the major reaction product. *J. Assoc. Off. Anal. Chem.,* **57**, 626–31.

Lembhe, A. F., Ranganathan, B., Ramana Rao, M. V. & Krishna Rao, L. (1981). Aflatoxin production in khoa. *J. Food Prot.,* **44**, 137–8.

Li, C. F. & Bradley, R. L., Jr (1969). Degradation of chlorinated hydrocarbon pesticides in milk and butter oil by ultraviolet energy. *J. Dairy Sci.,* **52**, 27–30.

Masimango, N., Remacle, J. & Ramaut, J. L. (1978). The role of adsorption in the elimination of aflatoxin B_1 from contaminated media. *Eur. J. Appl. Microbiol. Biotechnol.,* **6**, 101–5.

McKinney, J. D. & Cavanagh, G. C. (1977). Extraction of 'bound' aflatoxins. *Zeszyty Problemowe Postepow Nauk Rolniczych,* **189**, 247–54.

McKinney, J. D., Cavanagh, G. C., Bell, J. T., Hoversland, A. S., Nelson, D. M., Pearson, J. & Selkirk, R. J. (1973). Effect of ammoniation on aflatoxins in rations fed lactating cows. *J. Am. Oil Chem. Soc.,* **50**, 79–84.

Miller, M., Kiermeier, F., Weiss, G. & Klostermeyer, H. (1980). Aflatoxin M_1 in milk and milk products. *Z. Lebensm. Unters. Forsch.,* **171**, 20–3 (in German).

Moerck, K. E., McElfresh, P., Wohlman, A. & Hilton, B. W. (1980). Aflatoxin destruction in corn using sodium bisulfite, sodium hydroxide, and aqueous ammonia. *J. Food Prot.,* **43**, 571–4.

Odajima, T. (1981). Oxidative destruction of the microbial metabolite aflatoxin by the myeloperoxidase–hydrogen peroxide–chloride system. *Arch. Oral Biol.,* **26**, 339–40.

Parker, W. A. & Melnick, D. (1966). Absence of aflatoxin from refined vegetable oils. *J. Amer. Oil Chem. Soc.,* **43**, 635–8.

Patel, P. M., Netke, S. P., Gupta, D. S. & Dabadghao, A. K. (1981). Note on the effect of processing milk into khoa on aflatoxin M_1 content. *Indian J. Anim. Sci.,* **51**, 791–2.

Polzhofer, K. P. (1977). Thermal stability of Aflatoxin M_1. *Z. Lebensm. Unters. Forsch.,* **164**, 80–1 (in German).

Pons, W. A., Jr, Cucullu, A. F., Lee, L. S., Janssen, H. J. & Goldblatt, L. A. (1972). Kinetic study of acid-catalyzed conversion of aflatoxins B_1 and G_1 to B_{2a} and G_{2a}. *J. Amer. Oil Chem. Soc.,* **49**, 124–8.

Price, R. L., Jorgensen, K. V. & Billotte, M. (1981). Citrus artifact interference in aflatoxin M_1 determination in milk. *J. Assoc. Off. Anal. Chem.,* **64**, 1383–5.

Purchase, I. F. H. & Steyn, M. (1967). Estimation of aflatoxin M in milk. *J. Assoc. Off. Anal. Chem.,* **50**, 363–6.

Purchase, I. F. H., Steyn, M., Rinsma, R. & Tustin, R. C. (1972). Reduction of aflatoxin M content of milk by processing. *Food Cosmet. Toxicol.,* **10**, 383–7.

Reinbold, G. W. (1963). *Italian Cheese Varieties.* Chas. Pfizer and Co., Inc., New York, NY, USA.

Reiter, B. & Härnulv, G. (1984). Lactoperoxidase antibacterial system: natural occurrence, biological functions and practical applications. *J. Food Prot.,* **47**, 724–32.

Roberts, A. C. & McWeeny, D. J. (1972). The use of sulfur dioxide in the food industry, a review. *J. Food Technol.,* **7**, 221–38.

Schroeter, L. C. (1963). Kinetics of air oxidation of sulfurous acid salts. *J. Pharm. Sci.,* **52**, 559–63.

Sreenivasamurthy, V., Parpia, H. A. B., Srikanta, S. & Shankarmurti, A. (1967). Detoxification of aflatoxin in peanut meal by hydrogen peroxide. *J. Assoc. Off. Anal. Chem.,* **50**, 350–4.

Stoloff, L. (1980). Aflatoxin M_1 in perspective *J. Food Prot.,* **43**, 226–30.

Stoloff, L., Trucksess, M., Hardin, N., Francis, O. J., Hayes, J. R., Polan, C. E. & Campbell, T. C. (1975). Stability of aflatoxin M in milk. *J. Dairy Sci.,* **58**, 1789–93.

Stubblefield, R. D. & Shannon, G. M. (1974). Aflatoxin M_1: analysis in dairy products and distribution in dairy foods made from artificially contaminated milk. *J. Assoc. Off. Anal. Chem.,* **57**, 847–51.

Trager, W. & Stoloff, L. (1967). Possible reactions for aflatoxin detoxification. *J. Agr. Food Chem.,* **15**, 679–81.

Van Egmond, H. P. (1983). Mycotoxins in dairy products *Food Chem.,* **11**, 289–307.

Van Egmond, H. P. & Paulsch, W. E. (1986). Mycotoxins in milk and dairy products. *Neth. Milk Dairy J.,* **40**, 175–88.

Van Egmond, H. P., Paulsch, W. E., Veringa, H. A. & Schuller, P. L. (1977). The effect of processing on the aflatoxin M_1 content of milk and milk products. *Arch. Inst. Pasteur (Tunis),* **3–4,** 381–90.

Wei, R. & Chu, F. S. (1973). Aflatoxin–solvent interactions induced by ultraviolet light. *J. Assoc. Off. Anal. Chem.,* **56,** 1425–30.

Wiseman, D. W. & Marth, E. H. (1983a). Behavior of aflatoxin M_1 in yogurt, buttermilk and kefir. *J. Food Prot.,* **46,** 115–18.

Wiseman, D. W. & Marth, E. H. (1983b). Heat and acid stability of aflatoxin M_1 in naturally and artificially contaminated milk. *Milchwissenschaft,* **38,** 464–6.

Wiseman, D. W. & Marth, E. H. (1983c). Stability of aflatoxin M_1 during manufacture and storage of a butter-like spread, non-fat dried milk and dried buttermilk. *J. Food Prot.,* **46,** 633–6.

Wiseman, D. W. & Marth, E. H. (1983d). Stability of aflatoxin M_1 during manufacture and storage of ice-cream and sherbet. *Z. Lebensm. Unters. Forsch.,* **177,** 22–4.

Wiseman, D. W., Applebaum, R. S., Brackett, R. E. & Marth, E. H. (1983). Distribution and resistance to pasteurization of aflatoxin M_1 in naturally contaminated whole milk, cream and skim milk. *J. Food Prot.,* **46,** 530–2.

Wogan, G. N. (1964). Experimental toxicity and carcinogenicity of aflatoxins. In *Mycotoxins in Foodstuffs,* ed. G. N. Wogan. MIT Press, Cambridge, MA, USA, pp. 163–73.

Wogan, G. N. (1977). Mode of action of aflatoxins. In *Mycotoxins in Human and Animal Health,* ed. J. V. Rodricks, C. W. Hesseltine & M. A. Mehlman. Pathotox, Park Forest South, IL, USA, pp. 29–36.

Yousef, A. E. & Marth, E. H. (1985a). Degradation of aflatoxin M_1 in milk by ultraviolet energy. *J. Food Prot.,* **48,** 697–8.

Yousef, A. E. & Marth, E. H. (1985b). Rapid reverse phase liquid chromatographic determination of aflatoxin M_1 in milk. *J. Assoc. Off. Anal. Chem.,* **68,** 462–5.

Yousef, A. E. & Marth, E. H. (1986). Use of ultraviolet energy to degrade aflatoxin M_1 in raw or heated milk with and without added peroxide. *J. Dairy Sci.,* **69,** 2243–7.

Yousef, A. E. & Marth, E. H. (1987). Kinetics of interaction of aflatoxin M_1 in aqueous solutions irradiated with ultraviolet energy. *J. Agr. Food Chem.,* **35,** 785–9.

Chapter 6

Toxic Metabolites from Fungal Cheese Starter Cultures (*Penicillium camemberti* and *Penicillium roqueforti*)

G. Engel & M. Teuber

Institut für Mikrobiologie, Bundesanstalt für Milchforschung, Kiel, FRG

1. Introduction.. 163
   1.1 Fungal processed cheeses
   1.2 Moulds in cheese manufacture
   1.3 Potentially toxic metabolites
   1.4 Absence of carcinogenic mycotoxins

2. Toxic Metabolites Identified *in vitro* 168
   2.1 Methods of detection
     2.2 *P. camemberti*
     2.3 *P. roqueforti*

3. Toxic Metabolites Identified in Cheese.......................... 178
   3.1 Methods of detection
   3.2 Experimental cheeses
   3.3 Commercial cheeses

4. Prevention of Toxin Formation and Risk Assessment........................... 184
   4.1 Use of non-toxic or low toxic strains
  4.2 Ripening and storage conditions minimizing toxin production
     4.3 Health risk

5. Summary 187

1 INTRODUCTION

1.1 Fungal Processed Cheeses

Mould ripened cheeses are the blue veined cheeses and such with white surface mould. The most important blue veined cheeses are the Roquefort and Bleu de Bresse (France), Danablue (Denmark), Edelpilzkäse (Germany), Gorgonzola (Italy), and Stilton (Great

163

Britain), the prominent white surface mould cheeses are Camembert, Brie and Coulommiers.

In early cheesemaking, the moulds found their way into the milk and cheese by contamination from air and equipment. Today, commercially prepared mould spores powders or suspensions are either sprayed onto the surface of cheese, mixed with the curd before pressing or blended with the cheese-milk.

Fungi isolated from the surface or the interior of mould ripened cheese have been given different designations. In his elaborate analysis of *Penicillium* species, Pitt (1979) cleared up this confusing situation by assigning the different moulds used in cheese technology to the two species *Penicillium roqueforti* and *Penicillium camemberti* (see Table 1).

1.2 Moulds in Cheese Manufacture

1.2.1 *Penicillium camemberti*

Cheese varieties with surface mould are made using the so-called white mould cultures or Camembert cheese cultures. Here *P. camemberti* strains are used. According to Samson *et al.* (1977), these white coloured strains do not differ in their morphological and physiological properties from the green to greyish-blue coloured strains of *P. camemberti* cultures. For cheesemaking, this white type is used today, because a

TABLE 1
Synonyms of *P. camemberti* and *P. roqueforti* (Pitt, 1979)

Synonyms for P. camemberti *Thom*	*Synonyms for* P. roqueforti *Thom*
P. rogeri Wehmer apud Lafar	*P. aromaticum casei* Sopp
P. caseicolum Bain.	*P. vesiculosum* Bain.
P. camemberti var. *rogeri* Thom	*P. roqueforti* var. *weidemannii* Westling
P. epsteinii Lindau	*P. conservandi* Nov.
P. album Epstein	*P. casei* Staub
P. biforme Thom	*P. atroviride* Sopp
P. camemberti Sopp.	*P. roqueforti* Sopp
P. candidum Roger apud Biourge	*P. virescens* Sopp
P. paecilomyceforme von Szilvinyi	*P. stilton* Biourge
	P. biourgei Arnaudi
	P. suavelons Biourge
	P. gorgonzolae Weidemann apud Biourge
	P. weidemannii (Westling) Biourge

white fungal mycelium imparts an aesthetic appearance to the cheese. Colonies of such strains grow slowly on Czapek-agar at 25°C, attain a diameter of 2–3·5 cm within two weeks, consist of a raised floccose aerial mycelium usually up to 1 cm in height and remain white. Odour is mouldy or not pronounced. Their main functions in or on cheese are:

1. Protection from undesirable fungal infections by rapid formation of a mould layer on the cheese surface.
2. Proteolytic degradation of casein resulting in modification of cheese texture in the course of ripening.
3. Hydrolysis of triglycerides by lipolytic enzymes, which is associated with formation of aromatic substances.

1.2.2 *Penicillium roqueforti*

For the preparation of cheeses ripened in the interior under the influence of bluish to greenish moulds, and known under the collective term 'blue veined cheese', cultures of *P. roqueforti* are used. Colonies of such strains grow rapidly on Czapek-agar at 25°C, attain a diameter of 4–5 cm within two weeks, consist of a dense felt of erect conidiophores, and become more woolly (lanose) in some other cultures with production of aerial vegetative mycelium. The colour is blue-green, becoming darker later. The reverse side of the colonies is greenish, often changing to darker shades of green to black. Odour is mostly absent or not pronounced. During ripening of blue veined cheeses, intensive proteolysis takes place which is of importance for the texture and flavour of the cheese. In addition, *P. roqueforti* exhibits lipolytic activity. From the triglycerides, free fatty acids and ketones are formed which are important as aroma components (e.g. methylketones).

1.3 Potentially Toxic Metabolites

Mould cultures have been used for a long time without hesitation in cheesemaking until Gibel *et al.* (1971) published a report, suggesting that such cultures might present a hazard to the consumer's health by formation of toxic metabolites. It was described that a certain culture of *P. camemberti* seemed to produce a carcinogenic effect in rats after oral and subcutaneous application. In the meantime, it had been documented that *P. camemberti* or *P. roqueforti* strains are able to form several more or less toxic, well defined substances which are secondary metabolites originating from primary metabolites. In the synthesis of these toxic

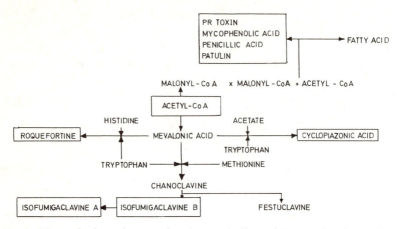

Fig. 1. Biosynthetic pathways of toxic metabolites of *P. roqueforti/camemberti* (Schoch, 1981).

secondary metabolites, acetyl-CoA (coenzyme A) plays a major role (see Fig. 1).

PR toxin (*Penicillium roqueforti* toxin), mycophenolic acid, penicillic acid and patulin are synthesized via malonyl-CoA, whilst roquefortine, cyclopiazonic acid and isofumigaclavine A and B, as well as festuclavine are derived from amino acids. Cyclopiazonic acid is the only toxin known so far which is formed by *P. camemberti*. The other mentioned metabolites may be produced by *P. roqueforti* strains.

1.4 Absence of Carcinogenic Mycotoxins

Many strains of *P. camemberti* or *P. roqueforti*, which are used as starter cultures or which have been isolated from cheese, have been examined for their ability to form a number of known mycotoxins which have harmful effects, e.g. aflatoxins (see Table 2).

The fungal cultures were incubated under suitable conditions in 2% yeast extract + 10–15% sucrose (YES) — a favourable nutrient medium for mycotoxin formation. Detection was mostly made physico-chemically after chromatographic separation of culture extracts.

The aflatoxins B and G belonging to the most carcinogenic substances as well as other mycotoxins were never found in extracts of these starter cultures nor in cheese made using these mould strains (Shih & Marth, 1969; Kiermeier & Groll, 1970). It is of some reassurance

TABLE 2

Studies on the production of aflatoxins and other mycotoxins by *P. camemberti* and *P. roqueforti* in yeast extract–sucrose medium

Mould strains	Number investigated	Mycotoxins investigated	Positive strains	Reference
P. camemberti	20	Aflatoxins B_1 and G_1, citrinin, ochratoxin A, zearalenone, penicillic acid, rubratoxin B, tremortin A	0	Mintzlaff & Machnik (1972)
P. roqueforti	7	Aflatoxins B_1 and G_1, citrinin, ochratoxin A, zearalenone, penicillic acid, rubratoxin B, tremortin A	0	Mintzlaff & Machnik (1972)
P. camemberti	41	Aflatoxins B_1, B_2 and G_1, sterigmatocystin, diacetoxyscirpenol, ochratoxin A, patulin, penicillic acid, citrinin, citreoviridin, rubratoxin B	0	Engel & von Milczewski (1977)
P. roqueforti	13	Aflatoxins B_1, B_2 and G_1, sterigmatocystin, diacetoxyscirpenol, ochratoxin A, patulin, penicillic acid, citrinin, citreoviridin, rubratoxin B	0	Engel & von Milczewski (1977)

to note that neither *P. camemberti* nor *P. roqueforti* form carcinogenic aflatoxins.

2 TOXIC METABOLITES IDENTIFIED *IN VITRO*

2.1 Methods of Detection

Detection of the formation of known toxic metabolites is mainly achieved by using chromatographic and spectrophotometric methods. The moulds are cultured in a medium suitable for toxin formation, e.g. yeast extract–sucrose nutrient solution, malt extract agar or cereals. The cultures are extracted with solvents such as chloroform, acetone, acetonitrile, methanol or mixtures of them. The extract components are separated chromatographically, mostly by thin layer chromatography (TLC), and the individual substances are detected by comparison with reference substances which have undergone the same chromatographic procedure. In Table 3, currently applied methods for the detection of toxic metabolites of *P. roqueforti* and *P. camemberti* are summarized. The most important identification and confirmation reactions are discussed in the paragraphs on the individual toxic metabolites.

Besides the physico-chemical methods which allow the detection of the type and concentration of known metabolites, various biological methods are suitable as screening tests for the detection of unknown toxic substances (see Table 4). As test organisms and systems trout, rats, mice, chick embryos, bacteria, *Artemia salina* shrimp larvae or human or animal tissue cultures have been used. In addition, the Ames-test (Ames *et al.*, 1973) for the detection of mutagenic effects, and long-term trials with animals to determine carcinogenic effects have been carried out.

2.2 *P. camemberti*

2.2.1 Cyclopiazonic acid

Cyclopiazonic acid is an optically active, colourless crystalline compound of the gross composition $C_{20}H_{20}N_2O_3$. Its structure (see Fig. 2) has been elucidated by Holzapfel (1968). Cyclopiazonic acid is visualized as a violet spot in daylight after spraying with Ehrlich reagent (Le Bars, 1979). The visual detection limit was $0·03 \, \mu g$, whilst the spectrodensitometric determination at 660 nm required at least $0·24 \, \mu g$ (Frevel, 1981).

TABLE 3

Summary of physico-chemical methods used for the detection of toxic metabolites of moulds including *P. camemberti* and *P. roqueforti*

Tested material	Separation/identification	Toxic metabolites	Reference
Pure toxins	TLC[a]	18 mycotoxins including patulin, penicillic acid	Scott *et al.* (1970)
Pure toxins	TLC	37 mycotoxins including mycophenolic acid, patulin, penicillic acid	Durackova *et al.* (1976)
Crude extracts from medium cultured with moulds:			
1. Yeast extract–sucrose	TLC	13 mycotoxins including patulin, penicillic acid	Engel & von Milczewski (1977)
2. Yeast extract–sucrose	TLC	Isofumigaclavine, roquefortine, PR toxin	Scott *et al.* (1977)
3. Malt agar	TLC	16 mycotoxins including cyclopiazonic acid, patulin, penicillic acid, PR toxin, roquefortine	Leistner & Eckardt (1979)
4. Maize	TLC	13 mycotoxins including cyclopiazonic acid, patulin, penicillic acid, roquefortine	Gorst-Allman & Steyn (1979)
5. Yeast extract–sucrose	TLC	Cyclopiazonic acid, mycophenolic acid, patulin, penicillic acid, PR toxin, roquefortine	Schoch *et al.* (1984a)

[a]TLC: thin layer chromatography.

TABLE 4

Summary of biological test systems, used for the detection of toxic metabolites from *P. camemberti* and *P. roqueforti*

Tested material	Biological test system	Reference
Pure substances: 16 mycotoxins including isofumiga- clavine, patulin, penicillic acid, PR toxin, roquefortine	Human and animal tissue culture (cytocidal effect, growth inhibition, morphological aberrations)	Lompe & v. Milczewski (1979)
Moulded rice: 3 *P. camemberti*, 1 *P. roqueforti*	Rainbow trout (increase in weight, appearance and weight of liver)	Frank *et al.* (1975)
Mould suspension: 3 *P. camemberti*, 1 *P. roqueforti*	Rat (subcutaneous application, observing for life, histological examination)	Frank *et al.* (1977)
Crude extracts from: Czapek's medium modified 2 *P. camemberti*, 29 *P. roqueforti*	Mouse (i.p. injection, acute toxicity), chicken embryo (inhibition of growth), cell culture (inhibition of multiplication of cells)	Lafont *et al.* (1976)

Yeast extract–sucrose 41 *P. camemberti*, 18 *P. roqueforti*	Tissue culture (histological examination)	Krusch *et al.* (1977)
Czapek's medium + glucose 19 fungi including 6 *P. roqueforti* from cheese	*Bacillus megatherium* (inhibition zone), chicken embryo (% mortality)	Moubasher *et al.* (1978)
Malt agar 1465 mould strains including 69 *P. camemberti,* 80 *P. roqueforti*	Brine shrimp larvae test with *Artemia salina* (mortality), chicken embryo (mortality)	Leistner & Eckardt (1979)
Czapek Dox + yeast extract + sucrose 18 *P. camemberti,* 8 *P. roqueforti*	Ames-test	Schoch *et al.* (1984*b*)

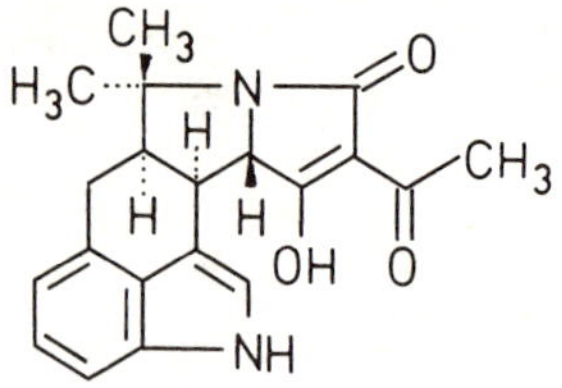

Fig. 2. Structure of cyclopiazonic acid.

Cyclopiazonic acid is acutely toxic to rats with an oral LD_{50} of 36 mg/kg body weight in males and 63 mg/kg in females (Purchase, 1971). In a study conducted by Morrissey *et al.* (1985) rats received cyclopiazonic acid at 0, 0·2, 2·0, 4·0 or 8·0 mg/kg body weight per day for 4 days. Clinical signs of toxicity were observed only in the two highest dose groups. Rats of these groups exhibited abnormal behaviour, diarrhoea, and other signs of toxicity after several days of dosing. Liver and spleen were more affected than other organs. In another experiment, rats were given cyclopiazonic acid at weekly doses of 0, 12 or 21 mg/kg body weight for 14 weeks. Males at the highest dose level initially showed mild growth retardation and 25% died suddenly during week four. No abnormal signs were observed in the surviving males or in any of the females throughout the experiment (Van Rensburg, 1984).

Intraperitoneal (i.p.) injections of cyclopiazonic acid produced hyperaesthesia and convulsions followed by death in rats in about 30 min. The LD_{50} is 2·3 mg/kg body weight (Purchase, 1971). Intraperitoneally applied cyclopiazonic acid also evoked a neurotoxic effect. Further target organs were the liver, spleen, pancreas and kidneys. The lesions observed were mostly found to be necroses. With i.p. administration, cyclopiazonic acid is rapidly distributed in the organism. If the toxin is orally administered, the symptoms either did not occur or were observed later. The reason may be that cyclopiazonic acid is barely soluble in water with a pH below 7 and is therefore absorbed much less rapidly (Schoch *et al.*, 1983).

Cyclopiazonic acid is the only known toxic metabolite product of *P. camemberti*. Since all white mould cultures of this type studied so far

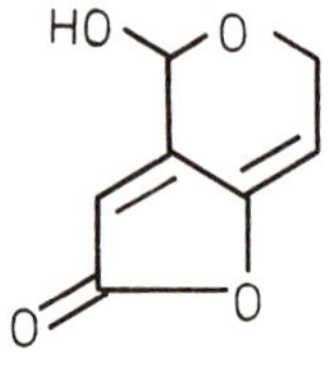

Fig. 3. Structure of patulin.

do form cyclopiazonic acid (Still *et al.*, 1978*a*; Le Bars, 1979; Engel, 1981*a*; Schoch *et al.*, 1984*a*) the conclusion has been drawn that *P. camemberti* is a consistent producer of this metabolite.

2.3 *P. roqueforti*

2.3.1 Patulin

Patulin is a colourless, crystalline compound which is optically inactive. The chemical structure (see Fig. 3) was elucidated by Dauben & Weisenborn (1949). After thin layer chromatography, patulin is detectable by spraying with *p*-anisaldehyde (visible light: faint brown, longwave UV light:yellow, detection limit 0·2 µg; Scott *et al.*, 1970), with 3-methyl-2-benzothiazolinone-hydrazone-hydrochloride (visible light: yellow, longwave UV light:yellowbrown fluorescent spot, detection limit 0·01 µg; Scott & Kennedy, 1973) or with *o*-phenylene-diamine (longwave UV light: green fluorescent spot, detection limit 0·01 µg; Meyer, 1975). Patulin is unstable in the presence of alkali (Brackett & Marth, 1979) and is inactivated by compounds containing SH-groups (Cavallito & Haskell, 1945; Geiger & Conn, 1945; Lieu & Bullerman, 1978). It has been shown that cysteine and glutathione react with patulin.

The following LD_{50} values of patulin were established for mice (M) and rats (R): orally 35 mg/kg body weight (M); subcutaneously (s.c.): 8–15 mg/kg (M), 15–25 mg/kg (R); intravenously (i.v.): 15–25 mg/kg (M), 25–50 mg/kg (R); intraperitoneally (i.p.): 15–30 mg/kg (M), 25 mg/kg (R) (Broom *et al.*, 1944; Scott, 1974). Dickens & Jones (1961) reported a carcinogenic activity of this substance, whilst Osswald *et al.* (1978) did not find a carcinogenic effect after feeding of patulin in a chronic trial with mice and rats. Patulin adducts formed with cysteine were found to have markedly lower toxicity values (Ciegler *et al.*, 1976; Lieu & Bullerman, 1978). Under defined conditions, patulin is formed by several *P. roqueforti* strains. These strains were, without exception, isolated as contaminating moulds from various foods. Whilst these strains, cultured as controls, formed patulin, the toxin was never found to be produced by strains used in the cheese industry (Harwig *et al.*, 1978; Olivigni & Bullerman, 1978; Leistner & Eckardt, 1979).

2.3.2 Penicillic acid

Penicillic acid is a colourless, crystalline compound and can exist in two tautomeric forms as a substituted γ-keto acid or as a γ-hydroxylactone

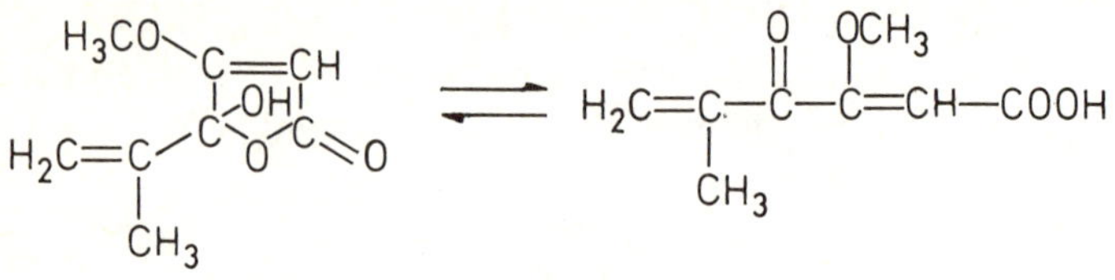

Fig. 4. Structure of penicillic acid.

(see Fig. 4; Birkinshaw *et al.*, 1936). The toxin is detectable after TLC by spraying with *p*-anisaldehyde (visible light: green, long wave UV light: blue fluorescence with a detection limit of 0·01 μg; Scott *et al.*, 1970) or exposure to ammonia vapour for 3 min (fluorescence: excitation at 350 nm, emission at 440 nm; Ciegler & Kurtzman, 1970). Like patulin, penicillic acid is inactivated by SH-groups, e.g. by cysteine and glutathione (Geiger & Conn, 1945; Lieu & Bullerman, 1978).

The following LD_{50} values were established for mice orally: 600 mg/ kg body weight; s.c.: 110 mg/kg; i.v.: 250 mg/kg; i.p.: 70 mg/kg (Murnaghan, 1946).

Dickens & Jones (1961, 1965) have extensively studied the possible carcinogenic action of lactones including penicillic acid. They observed that subcutaneous injections of 1·0 mg dosed twice weekly produced transplantable tumours after 64 weeks in all rats.

Formation of penicillic acid by *P. roqueforti* strains has been evidenced repeatedly. Penicillic acid was found in two out of six strains of *P. roqueforti* isolated from blue cheese (Moubasher *et al.*, 1978). Olivigni & Bullerman (1978) reported the production of penicillic acid by an atypical strain of *P. roqueforti* isolated from Cheddar cheese.

2.3.3 *Penicillium roqueforti* toxin (PR toxin)

Penicillium roqueforti toxin (PR toxin) is a colourless crystalline compound and is detected after TLC under UV light as a dark blue spot. The fluorescent colour changes to grey green after brief exposure to UV light and becomes visible in ordinary light as a yellow spot. The toxin can also be visualized on TLC plates after spraying with 50% sulphuric acid. It appears immediately as a yellow spot which changes into yellowish-brown after charring at 230°C for 2 min (Wei *et al.*, 1973). The detection limit is 0·15 μg after spraying with 50% sulphuric acid and heating at 10°C for 10 min (Engel, 1979). The chemical structure was elucidated by Wei *et al.* (1975) and is represented in Fig. 5.

The acute toxic data are summarized in Table 5. PR toxin inhibited

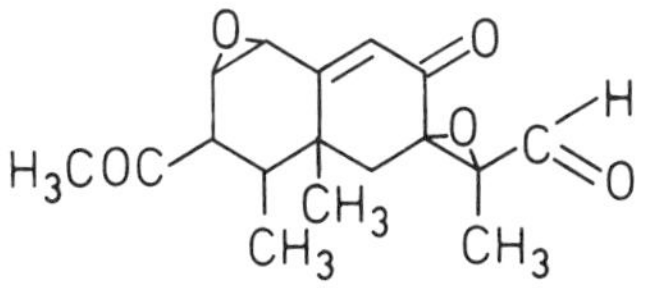

Fig. 5. Structure of PR toxin.

protein synthesis (Moulé *et al.*, 1978). The toxin was inactive in *Salmonella typhimurium* (Ames-test) and the sister-chromatid exchange assay (Moulé *et al.*, 1981). It is not yet known whether PR toxin is carcinogenic. PR toxin is unstable if added to blue cheese and to solvent extracts (methanol–water and chloroform) of blue cheese. It reacts with neutral and basic amino acids to form PR-imines. The toxicity of PR-imines is much lower than that of PR toxin (Table 5). When PR-imines are added to blue cheese they are unstable (Scott & Kanhere, 1979).

The ability of *P. roqueforti* strains to form PR toxin in synthetic and semi-synthetic media has been described in numerous studies. Orth (1976) tested one blue cheese starter and 11 isolates of *P. roqueforti* from different foodstuffs for the production of PR toxin. Three strains were able to produce the toxin (310–960 μg/kg) on rice at 15°C and sometimes at 30°C. Scott *et al.* (1977) showed that four strains of *P. roqueforti* isolated from blue cheese produced PR toxin. Wei & Liu (1978) tested

TABLE 5
Toxicity of PR toxin in rats and mice

Metabolite	Route	LD_{50} (mg/kg body weight)	Animal species	Reference
PR toxin	Oral	72–140	Mouse	Arnold *et al.* (1978)
	i.p.[a]	2–5	Mouse	Arnold *et al.* (1978)
	i.p.	5·8	Mouse	Chen *et al.* (1982)
	Oral	115	Rat	Wei *et al.* (1973)
	i.p.	11	Rat	Wei *et al.* (1973)
	i.p.	11·6	Rat	Chen *et al.* (1982)
	i.v.[b]	8·2	Rat	Chen *et al.* (1982)
PR toxin + Leucine	i.p.	100–200	Mouse	Arnold *et al.* (1978)
PR-imine	i.p.	100–200	Mouse	Arnold *et al.* (1978)

[a]Intraperitoneal.
[b]Intravenous.

different *P. roqueforti* strains from the American Type Culture Collection for the production of PR toxin. All strains were able to produce the toxin at 24°C after certain periods of incubation. The yields were correlated with the pH of the medium. Medina *et al.* (1985) observed PR toxin synthesis in YES medium by strains isolated from Cabrales Blue Cheese. However, they also observed that toxin formation decreased markedly in the presence of lactose or Na-lactate.

2.3.4 Mycophenolic acid

Mycophenolic acid was isolated from *Penicillium stoloniferum* (Alsberg & Black, 1913). It consists of colourless crystals. The structure (Fig. 6) was elucidated by Birkinshaw *et al.* (1948) and confirmed by Birch *et al.* (1958). Today this substance is mainly detected using TLC. Chromatograms are exposed to ammonia and observed under UV light: excitation at 360 nm, emission at 435 nm. The blue fluorescent derivative is unstable, losing 25 and 40% of its intensity after 5 and 10 min respectively (Lafont *et al.*, 1979*a*; Engel, 1981*b*).

Mycophenolic acid has antibiotic properties and the following LD_{50} values for rats (R) and mice (M): orally: 2500 mg/kg body weight (M), 700 mg/kg (R); i.v.: 550 mg/kg (M); 450 mg/kg (R) (Wilson, 1971).

Subacute toxicity was tested in the rabbit, monkey and rat. Oral doses of mycophenolic acid up to 320 mg/kg body weight for one year were given daily to rabbits (5 days/week). There were no apparent signs of toxicity (Adams *et al.*, 1975). Monkeys receiving daily 150 mg/kg body weight developed abdominal colic, loss of weight, and bloody diarrhoea after two weeks of feeding. These signs disappeared upon withdrawal of mycophenolic acid (Wilson, 1971). Rats given daily 30 mg/kg body weight became lethargic and pale, lost weight and were dead within nine weeks (Carter *et al.*, 1969).

Mycophenolic acid induced mutations and chromosome aberrations in a mouse mammary cell line (Umeda *et al.*, 1977), but was not mutagenic in *Salmonella* systems (Wehner *et al.*, 1978).

Studies by Lafont *et al.* (1979*a*) showed that all investigated 16 strains of *P. roqueforti* produced mycophenolic acid *in vitro* in amounts of

Fig. 6. Structure of mycophenolic acid.

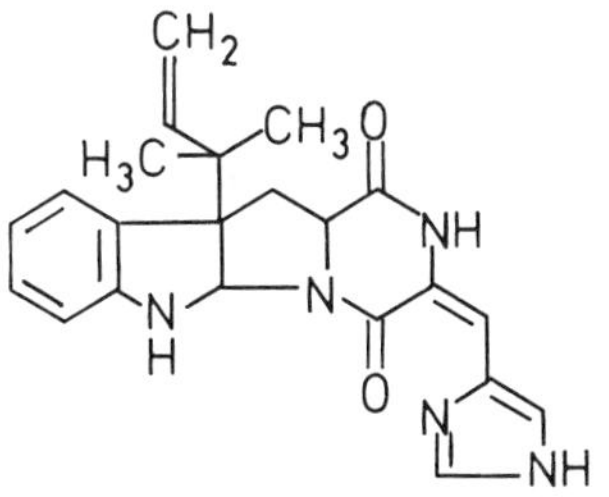

Fig. 7. Structure of roquefortine.

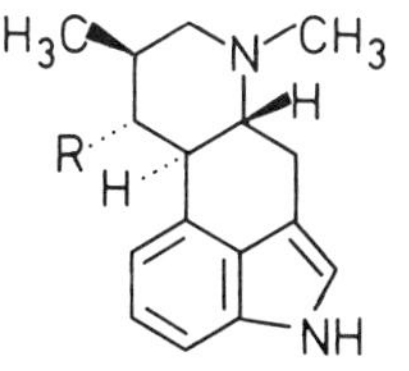

ISOFUMIGACLAVINE A (R= OCOCH₃)
ISOFUMIGACLAVINE B (R= OH)

Fig. 8. Structure of isofumigaclavines.

0·8–4 mg/g of dry culture. Best production occurred at 15°C after 10 days of incubation. Twenty of 80 strains of *P. roqueforti* were able to produce up to 600 mg of mycophenolic acid per litre in YES medium containing 5% sucrose (Engel *et al.*, 1982).

2.3.5 Roquefortine and isofumigaclavines

Ohmomo *et al.* (1975) and Scott *et al.* (1976) isolated and characterized different colourless crystalline alkaloids from *P. roqueforti* cultures. The most important of them are roquefortine and the isofumigaclavines A and B (Figs 7 and 8).

These substances are detected using TLC. TLC plates are sprayed with 50% sulphuric acid and heated at 110°C for 10 min. Roquefortine forms a light blue-grey spot with a detection limit of 0·15 μg (Engel, 1979) and isofumigaclavine A and B appear as mauve spots (Scott & Kennedy, 1976).

As to the biological activity of roquefortine, only a few data are available. The LD_{50} is 169–189 mg/kg body weight after i.p. administration to mice (Arnold *et al.*, 1978), other LD_{50} values are not known.

Observed neurotoxic properties attributed to roquefortine by C. Frayssinet and C. Frayssinet (cited by Scott *et al.*, 1976) could not be confirmed in similar experiments (Arnold *et al.*, 1978). Roquefortine tested for mutagenic activities in the Ames-test gave a negative result (Schoch *et al.*, 1983).

In experimental animals, isofumigaclavine A showed a variety of weak pharmacological actions, such as muscle relaxant, antidepressant, and local anaesthetic effects. Its LD_{50} in mice after i.p. administration however was 340 mg/kg body weight (Ohmomo *et al.*, 1975).

Formation of roquefortine by *P. roqueforti* cultures has been amply demonstrated. Engel & Teuber (1978) examined 40 *P. roqueforti* strains, isolated from blue cheese or used as starter cultures, for their ability to form roquefortine. Thirty-two of these cultures were found to be positive. Schoch *et al.* (1983) reported the formation of this compound by all of the six cheese factory strains studied, the concentration determined being dependent on both the temperature, time and the quantity of mycelium formed. No corresponding studies on isofumiga-clavines outside of cheeses are available.

3 TOXIC METABOLITES IDENTIFIED IN CHEESE

3.1 Methods of Detection

Various physico-chemical methods have been developed to detect toxic fungal metabolic products in cheese. These are mainly thin layer chromatographic methods summarized in Table 6 together with the corresponding detection limits.

In general, the cheese is extracted with chloroform, methanol, acetone, acetonitrile, ethyl acetate or mixtures of these solvents, sometimes in combination with water. Fats in the extracts are removed and separation of toxins from matrix components is achieved by TLC (often two-dimensional). Identification of the TLC spots usually takes place after derivatization. High performance liquid chromatography (HPLC) and gas chromatography (GC) usually require further time-consuming cleanup steps, so that these methods have rarely been applied. If only one substance is to be detected, it is possible to design the detection method especially for that substance. As a result, lower detection limits may be obtained compared with multi-detection methods, which are usually compromise procedures.

Animal tests have also been performed. Frank *et al.* (1977) and Schoch *et al.* (1984c) fed rats and mice with mould mycelium and samples of commercial blue, Camembert and Brie cheeses and examined the following parameters: life time, development of body weight, organ weight, histological examination of organs, and blood plasma enzymes.

3.2 Experimental Cheeses

Several investigators have studied the formation of toxic metabolites by *P. roqueforti* in cheese. Olivigni & Bullerman (1977) obtained patulin

TABLE 6
Summary of physico-chemical methods used for the detection of toxic metabolites of *P. camemberti* and *P. roqueforti* in cheese

Extractant	Separation/ identification	Toxic metabolite (detection limit, mg/kg)	Reference
Chloroform/methanol	TLC	Roquefortine (0·03), isofumigaclavine A, B (0·015)	Scott & Kennedy (1976)
Acetonitrile/chloroform	TLC	10 mycotoxins including patulin (0·3), penicillic acid (0·2)	Engel (1978)
Methanol/acetone	LC/TLC	6 mycotoxins including mycophenolic acid (0·02), patulin (0·02), penicillic acid (0·03)	Siriwardana & Lafont (1979)
Chloroform	LC/TLC	PR toxin (0·2)	Engel & Prokopek (1979)
Chloroform	TLC, two-dimensional	Cyclopiazonic acid (0·02)	Le Bars (1979), Frevel (1981)
Ethyl acetate	LC/HPLC	Roquefortine (0·016)	Ware *et al.* (1980)
Chloroform/methanol	TLC	Cyclopiazonic acid (0·02), mycophenolic acid (0·02), PR toxin (0·01), patulin (0·01), penicillic acid (0·02), roquefortine (0·03)	Schoch *et al.* (1984a)

and penicillic acid production at 5, 12, and 25°C on laboratory media, when inoculating with an atypical isolate of *P. roqueforti*, capable of producing both patulin and penicillic acid. These toxins were not produced however on Cheddar or Swiss cheeses. For the production of blue cheese, Engel and Prokopek (1980) used three *P. roqueforti* strains which had been isolated as contaminants from meat products and Cheddar cheese and which could form, under defined conditions, penicillic acid or patulin. Ripening was at 8°C and 15°C. Neither patulin nor penicillic acid were found in the cheeses at any time during the cheese ripening period (30–80 days), although under otherwise similar conditions marked toxin amounts of patulin up to 340 mg/litre and penicillic acid up to 600 mg/litre YES-medium were produced. According to Stott & Bullerman (1975) cheese is an unfavourable medium for patulin formation, which is apparently generally true for carbohydrate-poor and protein-rich substrates.

Studies have shown that in such substrates containing also glutathione or cysteine, patulin and penicillic acid react with the sulphydryl-groups of these compounds. This is associated with formation of addition products or adducts. The conversion rates observed here increase with increasing pH values (Lieu & Bullerman, 1978). These reactions could take place in cheese, the more so as also here there is an increase in pH with advanced ripening. In addition, the pH of the medium itself is of importance for the stability of patulin. Brackett & Marth (1979) calculated a half-life of 64 h at pH 8·0 and of 1310 h at pH 6·0 for the disappearance of patulin in a buffered medium.

Further interesting studies on the instability of patulin in Cheddar cheese were made by Stott & Bullerman (1976). After addition to cheese, there was a significant decrease of the biologically and chemically detectable toxin quantities independent of the temperature (5 and 25°). If patulin and penicillic acid are formed in cheese and react with SH-groups, the question of how toxic these patulin and penicillic acid adducts are arises. Studies with chicken embryos have shown that the toxicity of a patulin cysteine adduct is less than 1% of that of pure patulin (Ciegler *et al.*, 1976). Similar results were obtained by Lindroth & Wright (1978) in experiments with *E. coli* and mice (LD_{50}). Lieu & Bullerman (1978) found that penicillic acid and patulin adducts with cysteine or glutathione were markedly less or not toxic to *Bacillus subtilis*, larvae of sea shrimps or 4-day-old chicken embryos in the tested amounts (10–150 μg), compared with the corresponding mycotoxins.

Engel & Prokopek (1979) produced blue cheese using four *P. roqueforti* strains isolated from different foods, that were able to form PR toxin on laboratory media. In such cheeses no PR toxin was found during a 100-day ripening period at 8°C, although under the same incubation conditions up to 400 mg PR toxin/litre YES were formed. The same authors found that PR toxin is unstable in cheese. After addition of PR toxin to blue cheese, in a concentration of 8 mg/kg, 30% of the added PR toxin was detected after immediate extraction, 9·6% after 1 h storage, 3·9% after 3 and 5 h and only 1·3% after 24 h. If the same quantity of toxin was added to a mixture consisting of butterfat and water, or casein, tryptically-digested peptone and water, 30% of the added PR toxin was found after 24 h in butterfat and only 12% in the casein-peptone mixture. Scott & Kanhere (1979) confirmed these results. They found that PR toxin reacts with neutral and basic amino acids under formation of PR-imines and that this process also takes place in blue cheese. Further, they found that these PR-imines formed are also unstable in cheese. Arnold *et al.* (1978) reported that the toxicity of such compounds, peritoneally administered to mice, is less than 2% of the toxicity of pure PR toxin (see Table 5).

It is therefore not to be expected that PR toxin occurs in cheese. Besides this reaction, the question is left unanswered whether or not PR toxin can actually be formed in cheese, a medium generally unsuitable for mycotoxin formation. In addition, the formation of PR toxin is prevented by insufficient air admission, high NaCl concentrations and increase in the pH value in the neutral range, according to Piva *et al.* (1976).

To obtain proof that mycophenolic acid is synthesized in blue cheese by *P. roqueforti* strains used as a starter culture, Engel *et al.* (1982) prepared blue cheeses with producing and nonproducing strains, isolated from cheese and starter cultures. It was shown that only strains able to produce mycophenolic acid in YES broth produced this compound in cheese under the conditions of industrial cheese fermentation. However, the level of mycophenolic acid in cheese was about 50 to 100 times lower than that in broth.

3.3 Commercial Cheeses

The only known toxic metabolic product of *P. camemberti* is cyclo-piazonic acid, which is formed *in vitro* by all strains studied so far. Since

TABLE 7
Cyclopiazonic acid reported in commercial cheese samples (Camembert, Brie)

Number of examined cheeses	Number of positive samples	Concentration ranges of cyclopiazonic acid (mg/kg)	Reference
21	9	5:0·1–0·3[a] 4:0·7–1·9[a]	Still et al. (1978a, b)
20	11	3:0·05–0·1[b] 5:0·1–0·2[b] 3:0·4; 1; 1·5[b]	Le Bars (1979)
11	11	0·06–0·29[a]	Frevel (1981)
14	3	0·08; 0·25; 0·37[a]	Schoch et al. (1983)

[a]In the whole cheese.
[b]Calculated for the cheese rind.

nothing is known as yet of any reaction of this substance in cheeses, it may be expected to occur also in commercial cheeses. Table 7 summarizes the results obtained by different authors in studies on mould ripened soft cheese. The table shows that cyclopiazonic acid has been found in a great number of commercial cheese samples. It is to be expected that in each Camembert or Brie cheese cyclopiazonic acid will be detectable, if the corresponding methods of detection will obtain still lower limits of detection. According to Le Bars (1979) the occurrence is limited to the cheese rind and is probably not in the cheese interior. In cheese which is not yet fully ripened and stored in the cold the values are below 0·5 mg/kg, calculated on the whole cheese. Inappropriate storage at too high temperatures may lead to a drastic increase to values of 5 mg/kg (Still et al., 1978b).

Of the toxic metabolic products formed by P. roqueforti, patulin, penicillic acid, PR toxin, mycophenolic acid, roquefortine and isofumiga-clavines have been detected in commercial blue cheeses (see Table 8). Mycophenolic acid was only formed in cheese, if the starter cultures used were able to form this toxin in YES medium. Roquefortine was present in all blue cheeses examined, the highest concentration (6·8 mg/kg) was found in a selected piece with a visually high mould content. Only a few studies have been made on isofumigaclavines in cheese. Isofumiga-clavine A could be determined in concentrations of up to 4·7 mg/kg and in several samples traces of isofumigaclavine B were detectable.

TABLE 8
Metabolites of *P. roqueforti* reported in commercial blue cheese samples

Metabolite	Number of examined cheeses	Number of positive samples	Concentration ranges (mg/kg) in analysed samples	Reference
Mycophenolic acid	100	38	0·02–15	Lafont *et al.* (1979*b*)
	32	4	0·25–5	Engel *et al.* (1982)
Roquefortine	16	16	0·06–6·8	Scott & Kennedy (1976)
	12	12	0·16–0·65	Ware *et al.* (1980)
	13	13	0·20–2·3	Schoch *et al.* (1983)
Isofumigaclavine A	16	13	0·02–4·7	Scott & Kennedy (1976)
Isofumigaclavine B	16	6	Traces[a]	Scott & Kennedy (1976)
PR toxin	13	0		Engel & Prokopek (1979)
	Various samples	0		Polonelli *et al.* (1978)

[a]Only qualitative, identity confirmed by mass spectroscopy.

4 PREVENTION OF TOXIN FORMATION AND
RISK ASSESSMENT

4.1 Use of Non-toxic or Low Toxic Strains

According to the present state of knowledge of a possible mycotoxin formation by *P. roqueforti* cultures, the risk of formation of patulin, penicillic acid, PR toxin and mycophenolic acid can be excluded by selection of suitable strains. So far, patulin and penicillic acid have only been formed by those *P. roqueforti* strains which have been isolated as contaminant moulds from different foods. In starter cultures used for the production of blue cheese in the cheese industry, neither patulin nor penicillic acid synthesis have been detected as yet. PR toxin and mycophenolic acid have been formed by some starter cultures in defined culture solutions. However, PR toxin has not been detected in cheese as yet. Formation of both PR toxin and mycophenolic acid in cheese can be prevented by using only those mould strains in the production of blue cheese which do not form these substances in nutrient solutions suitable for toxin formation. Since such strains with excellent properties for cheese production are available, they should be used exclusively. Incidentally isofumigaclavines can be produced by certain starter cultures of *P. roqueforti*.

The situation concerning roquefortine formation by *P. roqueforti* and cyclopiazonic acid by *P. camemberti* is quite different. Roquefortine is synthesized by most, cyclopiazonic acid by all, of the starter cultures tested so far. It is therefore not surprising that both metabolic products have been found in the commercial cheeses examined (see Tables 7 and 8). The concentrations found (roquefortine: 0·06–6·8 mg/kg; cyclopiazonic acid: 0·06–5·7 mg/kg) varied markedly, the noted differences were apparently not caused by cheese varieties. For both substances, the quantity of mycelium formed seems to be important for the final concentration, apart from a strain-specific formation rate. Both metabolic products were mainly found in the mycelium-containing parts of cheese. It might be possible to eventually restrict roquefortine or cyclopiazonic acid synthesis by selecting suitable strains with limited ability to form these substances.

4.2 Ripening and Storage Conditions Minimizing
Toxin Production

Besides substrate and culture strain, the conditions of ripening and those prevailing in the cheese during ripening play a role in the production of metabolic products. The pH value decreases at first because of lactic acid formation and subsequently increases again, favoured by protein degradation (NH_3 formation), the ammonia formation being sensorally detectable in overripe cheese. As a result of this increase in pH, toxins such as penicillic acid, patulin, PR toxin and cyclopiazonic acid become unstable. The prevailing temperatures at ripening also have an influence on toxin production. Temperatures of 8°C permit the formation of patulin, penicillic acid, mycophenolic acid and PR toxin in nutrient solutions (Engel & Prokopek, 1979, 1980; Engel *et al.*, 1982). Roquefortine is also reported to be formed during ripening in the cheese cellar at 15°C (Schoch *et al.*, 1983). Formation of cyclopiazonic acid seems to be relatively limited at low temperatures (Frevel, 1981). The cheeses studied were all taken from cooled storage places (10°C) and had concentrations of up to 0·18 mg/kg cyclopiazonic acid in the whole cheese. Still *et al.* (1978*a*) stored cheese at 8–10°C without finding detectable amounts of cyclopiazonic acid after 28 days. Only after an unusually long period of storage (44 days) were concentrations of cyclopiazonic acid at 0·6 mg/kg detectable. However, when the cheese was stored at room temperature up to 5·7 mg/kg was found. In summary, the toxic metabolites mycophenolic acid and roquefortine formed by *P. roqueforti* and cyclopiazonic acid formed by *P. camemberti* are found in cheese at the usual ripening temperatures, although concentrations in artificial nutrient media may be much higher. The risk of a serious increase in concentration exists mainly if the cheese is not stored continuously at the temperatures which are normally applied (6–10°C).

4.3 Health Risk

Finally the question arises to what extent consumption of mould ripened cheeses presents a risk to health because of the toxic metabolic products possibly present in them.

Presence of patulin and penicillic acid formed by *P. roqueforti* is not to be expected, because producers of patulin and penicillic acid have not been detected among cheese starter cultures used in the cheese industry, and if cultures that are potential producers of patulin and penicillic acid were to be used, these toxins would be converted in cheese into non-toxic adducts through reaction with SH-containing amino acids. PR toxin produced by individual *P. roqueforti* strains has been found only in artificial nutrient media and has, so far, never been detected in commercial cheeses for various reasons. Only relatively few starter cultures form this toxin and the conditions of cheese ripening are not suitable for the formation of this toxin. The toxin is not stable in cheese and the reaction products formed are relatively non-toxic. Mycophenolic acid production in cheese has been observed experimentally and its presence in commercial cheese has been confirmed. For a realistic evaluation of any toxicological risk associated with consumption of cheese that contains mycophenolic acid, it is useful to recall the main experimental results pertinent to humans. A daily oral application of 40–80 mg/kg body weight in the treatment of psoriasis for 12 weeks did not induce harmful effects (Jones *et al.*, 1973). Rhesus monkeys, however, developed hypoplastic anaemia and severe intestinal disorders at daily doses of 150 mg/kg for two weeks. In contrast, the monkeys were without adverse effects at 50 mg/kg/day (Carter *et al.*, 1969). In view of the low incidence of mycophenolic acid-containing cheese on the market and the low content of mycophenolic acid (maximum: 5 mg/kg), no immediate toxicological risk is envisaged. Moreover, starter cultures are available which do not produce mycophenolic acid. Other metabolic products of *P. roqueforti* such as roquefortine and isofumiga-clavines have been detected in commercial blue cheese. The toxicity of these compounds is relatively low, however, and it is indeed questionable whether or not these fungal metabolites should be regarded as mycotoxins.

The only known toxic metabolic product of *P. camemberti* is cyclopiazonic acid, which was found (mainly in the cheese rind) in concentrations up to 0·5 mg/kg (with an average of 0·15 mg/kg), if the cheese was ripened under normal conditions and subsequently stored. According to the toxicological data currently available as well as the consumption habits concerning these cheese varieties no risk to human health in reality exists. Cyclopiazonic acid doses eventually ingested by consumers are very low compared with the oral LD_{50} dose in rats.

On the basis of the knowledge gained, acute lesions by consumption of mould ripened cheese can be excluded. The effects of mycotoxins present in cheese after continuous ingestion for longer periods of time were studied in feeding trials with *P. roqueforti* and *P. camemberti* cultures and with the cheeses in which they were used. After ingestion by rats or rainbow trout (Frank *et al.*, 1975, 1977; Schoch *et al.*, 1984*b*) no increased tumour rates or other histological changes and no changes in blood chemical parameters were found, compared with the controls. The doses of the daily administered quantities of mould were equivalent to the amount of mould in 100 kg of cheese per human per day.

Considering all data available in the literature, the presence of individual moderately toxic metabolites of *P. camemberti* (cyclopiazonic acid) and *P. roqueforti* (roquefortine) in cheese can never be completely excluded. Further selection of strains towards those with poor ability to form the mentioned metabolites and careful handling of the cheese during ripening and storage will further reduce the risk of toxin formation. Animal studies have not given any indications that other potential toxic metabolites might be formed in cheese by the starter cultures. In conclusion, consumption of mould ripened cheeses is not associated with hazards to human health (Teuber & Engel, 1983).

5 SUMMARY

For the production of mould ripened cheeses certain strains of *P. roqueforti* (in the case of blue veined cheeses) and of *P. camemberti* (in the case of white surface mould) are used. *In vitro, P. roqueforti* strains are able to form several more or less toxic metabolites such as patulin, penicillic acid, PR toxin, mycophenolic acid and roquefortine, whilst *P. camemberti* is a consistent producer of cyclopiazonic acid. In cheeses manufactured with *P. roqueforti* and *P. camemberti* only mycophenolic acid, roquefortine and cyclopiazonic acid could be detected.

In commercial mould ripened cheeses maximum amounts of 0·5 mg cyclopiazonic acid, 6·5 mg roquefortine and 4·7 mg isofumigaclavine A per kg of cheese have been detected. The production of mycophenolic acid in cheese can be prevented, if strains of *P. roqueforti* which cannot produce this substance are used. The mentioned toxins occur in cheese only in low amounts, they are only weakly toxic and not carcinogenic. Therefore it is very unlikely that consumers are endangered by consumption of mould ripened cheeses. This view was confirmed by animal feeding experiments with rats. Mould ripened cheeses, mould cultures, and extracts of these were given in quantities equivalent to more than 1 kg of cheese per day per kg body weight without adverse toxic or carcinogenic effects. In conclusion, consumption of mould ripened cheeses does not generate health hazards for the consumer.

REFERENCES

Adams, E., Todd, G. & Gibson, W. (1975). Long-term toxicity study of mycophenolic acid in rabbits. *Toxicol. Appl. Pharmacol.,* **34**, 509–12.

Alsberg, C. J. & Black, O. F. (1913). Biochemical and toxicological investigation of *Penicillium stoloniferum*. *US Department of Agriculture, Bureau of Plant Industry Bull. No 270*, 42–7.

Ames, B. A., Lee, F. D. & Durston, W. E. (1973). An improved bacterial test system for the detection and classification of mutagens and carcinogens. *Proc. Nat. Acad. Sci. USA,* **70**, 782–6.

Arnold, D. L., Scott, P. M., McGuire, P. E., Harwig, H. & Neva, E. A. (1978). Acute toxicity studies on roquefortine and PR toxin, metabolites of *Penicillium roqueforti*, in the mouse. *Fd Cosmet. Toxicol.,* **16**, 369–71.

Birch, A. J., English, R. J., Massy-Westropp, R. A. & Smith, H. (1958). Studies in relation to biosynthesis. Part XV, Origin of terpenoid structures in mycelianamide and mycophenolic acid. *J. Chem. Soc.,* 369–75.

Birkinshaw, J. H., Oxford, A. E. & Raistrick, H. (1936). Studies in the biochemistry of microorganisms. Penicillic acid, a metabolic product of *Penicillium puberulum* Bainier and *P. cyclopium* Westling. *Biochem. J.,* **30**, 334–411.

Birkinshaw, J. H., Bracken, E., Morgan, E. N. & Raistrick, H. (1948). The molecular constitution of mycophenolic acid, a metabolic product of *Penicillium brevi-compactum* Dierckx. *Biochem. J.,* **43**, 216–23.

Brackett, R. E. & Marth, E. H. (1979). Stability of patulin at pH 6·0–6·8 and 25°C. *Z. Lebensm. Unters. Forschg.,* **169**, 92–4.

Broom, W. A., Bülbring, E., Chapman, C. J., Hampton, J. W., Thomson, A. M., Ungar, J., Wien, R. & Woolfe, G. (1944). The pharmacology of patulin. *Brit. J. exptl. Pathol.,* **25**, 195–207.

Carter, S. B., Franklin, T. J., Jones, D. F., Leonard, B. J., Mills, S. D., Turner, R. W. & Turner, W. B. (1969). Mycophenolic acid: anti-cancer compound with unusual properties. *Nature,* **233**, 848–50.

Cavallito, C. J. & Haskell, T. H. (1945). The mechanism of action of antibiotics. The reaction of unsaturated lactones with cysteine and related compounds. *J. Amer. Chem. Soc.,* **67**, 1991–4.

Chen, F. C., Chen, C. F. & Wei, R. D. (1982). Acute toxicity of PR toxin, a mycotoxin from *Penicillium roqueforti*. *Toxicon,* **20**, 433–41.

Ciegler, A. & Kurtzman, C. P. (1970). Fluorodensitometric assay of penicillic acid. *J. Chromatogr.,* **51**, 511–16.

Ciegler, A., Beckwith, A. C. & Jackson, L. K. (1976). Teratogenicity of patulin and patulin adducts with cysteine, *Appl. Environ. Microbiol.,* **31**, 664–7.

Dauben, J. & Weisenborn, F. L. (1949). The structure of patulin. *J. Amer. Chem. Soc.,* **71**, 3853.

Dickens, F. & Jones, H. E. H. (1961). Carcinogenic activity of a series of reactive lactones and related substances. *Brit. J. Cancer,* **15**, 85–100.

Dickens, F. & Jones, H. E. H. (1965). Further studies on the carcinogenic action of certain lactones and related substances in the rat and mouse. *Brit. J. Cancer,* **19**, 392–403.

Durackova, Z., Betina, V. & Nemec, P. (1976). Systematic analysis of mycotoxins by thin-layer chromatography. *J. Chromatogr.,* **116**, 141–54.

Engel, G. (1978). Formation of mycotoxins on Tilsit cheese. *Milchwissensch.,* **33**, 201–3 (in German).

Engel, G. (1979). UV-densitometric analysis of roquefortine and PR toxin. *J. Chromatogr.,* **170**, 288–91 (in German).

Engel, G. (1981*a*). Production of cyclopiazonic acid by *Penicillium camemberti* and *P. caseicolum. Annual Report of the Federal Dairy Research Centre Kiel,* B33–B34 (in German).

Engel, G. (1981*b*). Quantitative UV-densitometric and fluorodensitometric analysis of mycophenolic acid on thin layer plates. *J. Chromatogr.,* **207**, 430–4 (in German).

Engel, G. & von Milczewski, K. E. (1977). *Penicillium caseicolum, P. camemberti* and *P. roqueforti* and their harmlessness to human health: I. Physicochemical tests for the formation of known mycotoxins. *Milchwissensch.,* **32**, 517–20 (in German).

Engel, G. & Prokopek, D. (1979). No detection of *Penicillium roqueforti*-toxin in cheese. *Milchwissensch.,* **34**, 272–4 (in German).

Engel, G. & Prokopek, D. (1980). No detection of patulin and penicillic acid in cheese produced by *Penicillium roqueforti*-strains forming patulin and penicillic acid. *Milchwissensch.,* **35**, 218–20 (in German).

Engel, G. & Teuber, M. (1978). Simple aid for the identification of *Penicillium roqueforti* Thom: Growth in acetic acid. *Europ. J. Appl. Microbiol. Biotechnol.,* **6**, 107–11.

Engel, G., von Milczewski, K. E., Prokopek, D. & Teuber, M. (1982). Strain-specific synthesis of mycophenolic acid by *Penicillium roqueforti* in blue-veined cheese. *Appl. Environ. Microbiol.,* **43**, 1034–40.

Frank, J. K., Orth, R., Reichle, G. & Wunder, W. (1975). Feeding experiments with rainbow trout using Camembert and Roquefort starters. *Milchwissensch.,* **30**, 594–7 (in German).

Frank, J. K., Orth, R., Ivankowic, S., Kuhlmann, M. & Schmähl, D. (1977). Investigations on carcinogenic effects of *Penicillium caseicolum* and *P. roqueforti* in rats. *Experientia,* **33**, 515–16.

Frevel, H. J. (1981). Isolation, identification and detection of cyclopiazonic acid in Camembert-cheese. Diplomarbeit an der Christian-Albrechts-Universität, Kiel (in German).

Geiger, W. B. & Conn, J. E. (1945). The mechanism of antibiotic action of clavacin and penicillic acid. *J. Amer. Chem. Soc.,* **67**, 112–16.

Gibel, W., Wegner, K. & Wildner, G. P. (1971). Experimental investigations on the cancerogenicity of *Penicillium camemberti* var. *candidum. Archiv für Geschwulstforschung,* **38**, 1–6 (in German).

Gorst-Allmann, C. P. & Steyn, P. S. (1979). Screening methods for the detection of thirteen common mycotoxins. *J. Chromatogr.,* **175**, 325–31.

Harwig, J., Blanchfield, B. J. & Scott, P. M. (1978). Patulin production by *Penicillium roqueforti* Thom from grape. *Can. Inst. Food Sci. Technol. J.,* **11**, 149–51.

Holzapfel, C. W. (1968). The isolation and structure of cyclopiazonic acid, a toxic metabolite of *Penicillium cyclopium* Westling. *Tetrahedron,* **24**, 2101–19.

Jones, E. L., Epinette, W. W., Hackney, V. C., Manendez, L. & Frost, P. (1973). Treatment of psoriasis with oral mycophenolic acid. *J. Invest. Dermatol.,* **60**, 246.

Kiermeier, F. & Groll, D. (1970). About the formation of aflatoxin B_1 in cheese. *Z. Lebensm. Unters. Forschg.,* **143**, 81–9 (in German).

Krusch, U., Lompe, A., Engel, G. & von Milczewski, K. E. (1977). *Penicillium caseicolum, P. camemberti* and *P. roqueforti* and their harmlessness to human health. II. Biological testing of toxin formation ability on cell cultures. *Milchwissensch.,* **32**, 713–15 (in German).

Lafont, P., Lafont, J., Payen, J., Chang, E., Bertin, G. & Frayssinet, C. (1976). Toxin production by 50 strains of *Penicillium* used in the cheese industry. *Fd Cosmet. Toxicol.,* **14**, 137–9.

Lafont, P., Debeaupuis, J. P., Gaillardin, M. & Payen, J. (1979a). Production of mycophenolic acid by *Penicillium roqueforti* strains. *Appl. Environ. Microbiol.,* **37**, 365–8.

Lafont, P., Siriwardana, M. G., Combemale, J. & Lafont, J. (1979b). Mycophenolic acid in marketed cheeses. *Fd Cosmet. Toxicol.,* **17**, 147–9.

Le Bars, J. (1979). Cyclopiazonic acid production by *Penicillium camemberti* Thom and natural occurrence of this mycotoxin in cheese. *Appl. Environ. Microbiol.,* **38**, 1052–5.

Leistner, L. & Eckardt, C. (1979). Occurrence of toxicogenic Penicillia in meat products. *Fleischwirtsch.,* **59**, 1892–6 (in German).

Lieu, F. Y. & Bullerman, L. B. (1978). Binding of patulin and penicillic acid to glutathione and cysteine and toxicity of the resulting adducts. *Milchwissensch.,* **33**, 16–20.

Lindroth, S. & Wright, A. (1978). Comparison of the toxicities of patulin and patulin adducts formed with cysteine. *Appl. Environ. Microbiol.,* **35**, 1003–7.

Lompe, A. & von Milczewski, K. E. (1979). A cell-culture assay for the detection of mycotoxins. *Z. Lebensm. Unters. Forschg.,* **169**, 249–54 (in German).

Medina, M., Gaya, P. & Nunez, M. (1985). Production of PR toxin and roquefortine by *Penicillium roqueforti* isolates from cabrales blue cheese. *J. Fd Prot.,* **48**, 118–21.

Meyer, R. A. (1975). Ein neuer dünnschichtchromatographischer Nachweis für Patulin. *Die Nahrung,* **19**, K1–K2.

Mintzlaff, H. J. & Machnik, W. (1972). Untersuchungen über das Toxin-bildungsvermögen von *Penicillium caseicolum* and *Penicillium roqueforti*-Stämmen, die für die Hertstellung von verschiedenen Käsearten von Bedeutung sind. *Annual Report of the Federal Meat Research Centre (Kulmbach),* 151–2.

Morrissey, R. E., Norred, W. P., Cole, R. J. & Dorner, J. (1985). Toxicity of the mycotoxin cyclopiazonic acid to Sprague–Dawley rats. *Toxicol. Appl. Pharmacol.,* **77**, 94–107.

Moubasher, A. H., Abdel-Kader, M. J. A. & El-Kadi, J. A. (1978). Toxigenic fungi isolated from Roquefort cheese. *Mycopathologia,* **66**, 187–90.

Moulé, Y., Jemmali, M. & Darracq, N. (1978). Inhibition of protein synthesis by PR toxin, a mycotoxin from *Penicillium roqueforti*. *FEBS Letters,* **88**, 341–4.

Moulé, Y., Hermann, M. & Renault, G. (1981). Negative response of PR toxin in the *Salmonella typhimurium*/microsome test and sister-chromatid exchange assay. *Mutation Res.,* **89**, 203–7.

Murnaghan, M. F. (1946). The pharmacology of penicillic acid. *J. Pharmacol. Exp. Ther.,* **88**, 119–32.

Ohmomo, S., Sato, T., Utagawa, T. & Abe, M. (1975). Isolation of festuclavine and three new indole alkaloids, roquefortine A, B and C from the cultures of *Penicillium roqueforti. Agr. Bio. Chem.,* **39**, 1333–4.

Olivigni, F. J. & Bullerman, L. B. (1977). Simultaneous production of penicillic acid and patulin by a *Penicillium* species isolated from Cheddar cheese. *J. Fd Sci.,* **42**, 1654–65.

Olivigni, F. J. & Bullerman, L. B. (1978). Production of penicillic acid and patulin by an atypical *Penicillium roqueforti* isolate. *Appl. Environ. Microbiol.,* **35**, 435–8.

Orth, R. (1976). PR-toxin production of *Penicillium roqueforti* strains. *Z. Lebensm. Unters. Forschg.,* **160**, 131–6 (in German).

Osswald, H., Frank, H. K., Komitowski, D. & Winter, H. (1978). Long-term testing of patulin administered orally to Sprague-Dawley rats and Swiss mice. *Fd Cosmet. Toxicol.,* **16**, 243–7.

Pitt, J. I. (1979). *The Genus* Penicillium *and its Teleomorphic States* Eupenicillium *and* Talaromyces. Academic Press, London.

Piva, M., Guiraud, J., Crouzet, J. & Galzy, P. (1976). Influence des conditions de culture sur l'excrétion d'une mycotoxine par quelques souches de *Penicillium roqueforti. Le Lait,* **557**, 397–406.

Polonelli, L., Morace, G., Monache, F. D. & Samson, R. A. (1978). Studies on the PR toxin of *Penicillium roqueforti. Mycopathologia,* **66**, 99–104.

Purchase, I. F. H. (1971). The acute toxicity of the mycotoxin cyclopiazonic acid to rats. *Toxicol. Appl. Pharmacol.,* **18**, 114–23.

Samson, R. A., Eckardt, C. & Orth, R. (1977). The taxonomy of *Penicillium* species from fermented cheeses. *Antonie van Leeuwenhoek,* **43**, 341–50.

Schoch, U. (1981). Mycotoxins in mould-ripened cheese — a review. *Mitt. Geb. Lebensm. Hyg.,* **72**, 380–95 (in German).

Schoch, U., Lüthy, J. & Schlatter, C. (1983). Mycotoxins in mould-ripened cheese. *Mitt. Geb. Lebensm. Hyg.,* **74**, 50–9 (in German).

Schoch, U., Lüthy, J. & Schlatter, C. (1984a). Mycotoxins of *P. roqueforti* and *P. camemberti* in cheese. *Milchwissensch.,* **39**, 76–80 (in German).

Schoch, U., Lüthy, J. & Schlatter, C. (1984b). Mutagenicity testing of commercial *P. camemberti* and *P. roqueforti*-strains. *Z. Lebensm. Unters. Forschg.,* **178**, 351–5 (in German).

Schoch, U., Lüthy, J. & Schlatter, C. (1984c). Subcronic toxicity testing of mould ripened cheese. *Z. Lebensm. Unters. Forschg.,* **179**, 99–103 (in German).

Scott, P. M. (1974). Patulin. In *Mycotoxins*, ed. I. F. H. Purchase. Elsevier Scientific Publ. Co., Amsterdam, pp. 383–403.

Scott, P. M. & Kanhere, S. R. (1979). Instability of PR toxin in blue cheese. *J. Assoc. Off. Anal. Chem.,* **62**, 141–7.

Scott, P. M. & Kennedy, B. P. C. (1973). Improved method for the thin-layer chromatographic determination of patulin in apple juice. *J. Assoc. Off. Anal. Chem.,* **56**, 813–16.

Scott, P. M. & Kennedy, B. P. C. (1976). Analysis of blue cheese for roquefortine and other alkaloids from *Penicillium roqueforti. J. Agric. Fd Chem.,* **24**, 365–8.

Scott, P. M., Lawrence, J. W. & van Walbeek, W. (1970). Detection of mycotoxins by thin-layer chromatography: application to screening of fungal extracts. *Appl. Microbiol.,* **20**, 839–42.

Scott, P. M., Merrien, M. A. & Polonsky, J. (1976). Roquefortine and isofumiga-clavine A, metabolites from *Penicillium roqueforti. Experientia,* **32,** 140–2.

Scott, P. M., Kennedy, B. P. C., Harwig, J. & Blanchfied, B. J. (1977). Study of conditions for production of roquefortine and other metabolites of *Penicillium roqueforti. Appl. Environ. Microbiol.,* **33,** 249–53.

Shih, C. N. & Marth, E. H. (1969). Aflatoxins not recovered from commercial mold-ripened cheeses. *J. Dairy Sci.,* **52,** 1681–2.

Siriwardana, M. G. & Lafont, P. (1979). Determination of mycophenolic acid, penicillic acid, patulin, sterigmatocystin and aflatoxins in cheese. *J. Dairy Sci.,* **62,** 1145–8.

Still, P. E., Eckardt, C. & Leistner, L. (1978*a*). Bildung von Cyclopiazonsäure durch *Penicillium camemberti*-Isolate von Käse. *Die Fleischwirtsch.,* **5,** 876–7.

Still, P. E., Eckardt, C. & Leistner, L. (1978*b*). Annual Report of the Federal Meat Research Centre (Kulmbach) C30.

Stott, W. T. & Bullerman, L. B. (1975). Patulin: a mycotoxin of potential concern in foods. *J. Milk Fd Technol.,* **38,** 695–705.

Stott, W. T. & Bullerman, L. B. (1976). Instability of patulin in Cheddar cheese. *J. Fd Sci.,* **41,** 201–3.

Teuber, M. & Engel, G. (1983). Low risk of mycotoxin production in cheese. *Microbiologie–Aliments–Nutrition,* **1,** 193–7.

Umeda, U., Tsutsui, T. & Saito, M. (1977). Mutagenicity and inducibility of DNA single-strand breaks and chromosome abberations by various mycotoxins. *Gann,* **68,** 619–25.

Van Rensburg, S. J. (1984). Subacute toxicity of the mycotoxin cyclopiazonic acid. *Fd Chem. Toxicol.,* **22,** 993–8.

Ware, G. M., Thorpe, C. W. & Pohland, A. E. (1980). Determination of roque-fortine in blue cheese and blue cheese dressing by high pressure liquid chromatography with ultraviolet and electrochemical detectors. *J. Assoc. Off. Anal. Chem.,* **63,** 637–41.

Wehner, F. C., Thiel, P. G., van Rensburg, S. J. & Demasius, J. P. C. (1978). Mutagenicity to *Salmonella typhimurium* of some *Aspergillus* and *Penicillium* mycotoxins. *Mutation Res.,* **58,** 193–203.

Wei, R. D. & Liu, G. X. (1978). PR toxin production in different *Penicillium roqueforti* strains. *Appl. Environ. Microbiol.,* **35,** 797–9.

Wei, R. D., Still, P. E., Smalley, E. B., Schnoes, H. K. & Strong, F. M. (1973). Isolation and partial characterization of a mycotoxin from *Penicillium roqueforti. Appl. Microbiol.,* **25,** 111–14.

Wei, R. D., Schnoes, H. K., Hart, P. A. & Strong, F. M. (1975). The structure of PR toxin, a mycotoxin from *Penicillium roqueforti. Tetrahedron,* **31,** 109–114.

Wilson, B. J. (1971). Miscellaneous *Penicillium* toxins. In *Microbial Toxins, Vol. VI,* ed. A. Ciegler, S. Kadis & S. J. Ajl. Academic Press, New York, pp. 460–9.

Chapter 7

Mycotoxigenic Fungal Contaminants of Cheese and Other Dairy Products

P. M. Scott

Health and Welfare Canada, Health Protection Branch,
Bureau of Chemical Safety, Ottawa, Canada

1. Introduction 194
2. Mycotoxigenic Fungi Isolated from Cheese 196
    2.1 General
    2.2 *Penicillium* species
    2.3 *Aspergillus* species
    2.4 Other fungal genera isolated from cheese
    2.5 Conclusions
3. Ability of Fungi to Produce Mycotoxins on Cheese 214
    3.1 General
    3.2 *Penicillium* toxins
    3.3 *Aspergillus* toxins
    3.4 Prevention of fungal growth
    3.5 Conclusions
4. Penetration and Stability of Mycotoxins in Cheese 222
    4.1 General
    4.2 Migration of mycotoxins in cheese
    4.3 Practical conclusions
    4.4 Stability of mycotoxins in cheese
5. Natural Occurrence of Mycotoxins in Mouldy Cheese 227
    5.1 General
    5.2 *Penicillium* toxins
    5.3 *Aspergillus* toxins
    5.4 Conclusions
6. Other Dairy Products 234
    6.1 Fungi isolated from dairy products other than cheese
    6.2 Production of mycotoxins in dairy products other than cheese
    6.3 Natural occurrence of mycotoxins in dairy products other than cheese
    6.4 Conclusions
7. Analytical Methods for Mycotoxins with Potential to Contaminate Dairy Products Directly 240
    7.1 General
    7.2 Methods for *Penicillium* mycotoxins (multimycotoxin methods)
    7.3 Sterigmatocystin
    7.4 Aflatoxins B_1, B_2, G_1 and G_2
    7.5 β-nitropropionic acid
    7.6 Conclusions
8. Summary 244

193

1 INTRODUCTION

In the previous chapter (Chapter 6), the *Penicillium* species that were discussed were intentionally used for the purpose of ripening Camembert-type and blue cheeses. However, many extraneous moulds find cheese an excellent medium for growth and cheese can become mouldy during the ripening process and after cutting and slicing with unclean equipment in shops or at home. Moulds are usually considered as undesirable contaminants that may lead to complete spoilage of the cheese if it is stored long enough, even under refrigeration. Consumer complaints about mould in foods in the United Kingdom are 4000–6000 annually (Jarvis *et al.*, 1985); the data from 1970 showed cheese ranked fourth after bread, meat pies and confectionery (Jarvis, 1972). For some types of cheese, notably certain French cheeses, such as Saint-Nectaire, Tome de Savoie, and various goat's milk cheeses, it is considered desirable for a mouldy surface to develop naturally during ripening. As will be discussed in this chapter, the main genus of mould isolated from cheese is *Penicillium*. While it is true that many species of *Penicillium* produce the antibiotic penicillin (Korzybski *et al.*, 1967), such species appear to be rare as cheese contaminants. *P. chrysogenum* and *P. notatum* are two species (regarded as synonymous by Pitt & Hocking (1985)) that are known to produce penicillin and have been isolated from several cheeses (Delespaul *et al.*, 1973; Demirer, 1974; Galli & Zambrini, 1978; Northolt & Soentoro, 1979; Northolt *et al.*, 1980; Chapman *et al.*, 1983; Jarvis, 1983; Polonelli *et al.*, 1984). However, penicillin does not appear to have actually been looked for in a fungal isolate from cheese nor in cheese itself. Thus there is no evidence to support the commonly held opinion that because of penicillin mouldy cheese is good for one's health. Other antibiotics, such as mycophenolic acid and patulin now classified as mycotoxins, have been found in mouldy cheese (Section 5.2).

The questions of what fungi have been found as cheese contaminants, whether they can form mycotoxins, whether cheese is a good substrate for mycotoxin production, to what extent mycotoxins can penetrate into cheese, how stable they are in cheese, and how common mycotoxin occurrence is in cheese will be dealt with in this review. Additionally, the potential for mycotoxin occurrence in other dairy products due to direct fungal contamination as well as analytical methods for myco-toxins in cheese and other dairy products will be discussed. Several reviews on the topics of moulds and mycotoxins in cheese or dairy products in general have been published (Brandl, 1976; Moreau, 1976*a*, 1983; Bullerman, 1977*b*, 1981; Lück & Wehner, 1979; Josefsson, 1981; Kiermeier, 1981; Applebaum *et al.*, 1982; Tantaoui-Elaraki & Khabbazi, 1984; Pfleger, 1985); these include sections on direct fungal contamination. For background information on mycotoxins, the following compilations should be consulted (Cole & Cox, 1981; Betina, 1984; Ueno, 1985; Watson, 1985). Toxicologically, mycotoxins that have been found in cheese may be regarded as nephrotoxic (ochratoxin A, citrinin), teratogenic (ochratoxin A, aflatoxin B_1), neurotoxic/tremorgenic (penitrem A, α-cyclopiazonic acid), orally carcinogenic (aflatoxins B_1 and G_1, sterigmatocystin, ochratoxin A), carcinogenic by injection (patulin, penicillic acid), or toxic antibiotics (patulin, penicillic acid, mycophenolic acid, citrinin) (Betina, 1984; Ueno, 1985). Structural formulae of the more important mycotoxins found in cheese are shown in Fig. 1.

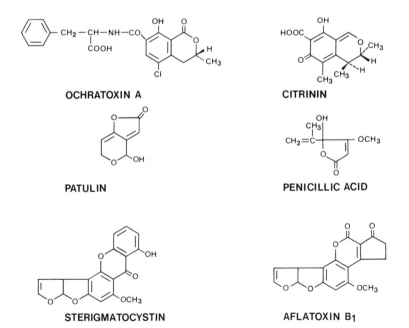

Fig. 1. Structural formulae of some mycotoxins that have been detected in cheese.

2 MYCOTOXIGENIC FUNGI ISOLATED FROM CHEESE

2.1 General

Cheese may be regarded as either non-visibly or visibly mouldy and considerable effort has gone into isolating, identifying and testing the toxicity of fungi from both types of cheese. If visible mould is trimmed from mouldy cheese, the remainder can be expected to have a much higher mould count than cheese that had no visible mould originally, which tends to have low mould counts (Bullerman, 1980). Prolonged storage can then permit mould growth, particularly if the package is opened. It is thus important to know what kind of fungi are present in all cheeses, both visibly and non-visibly mouldy. A general summary is given in Table 1 of studies where incidence data on the distribution of various fungi on cheese were reported. These studies clearly show that *Penicillium* species predominate in the fungal mycoflora of these cheeses. The highest incidences of *Aspergillus* species have been found on Gouda and Edam cheese stored in warehouses in The Netherlands (Northolt & Soentoro, 1979; Northolt *et al.*, 1980; Northolt & Van Egmond, 1982*a*), on Provolone cheese (Galli & Zambrini, 1978), and in Egyptian cheeses (El-Bassiony *et al.*, 1980; Mahmoud *et al.*, 1983).

2.2 *Penicillium* Species

2.2.1 General
The mould mycoflora of cheese are mainly *Penicillium* species and, furthermore, moulds that develop on cheese during refrigerated storage are almost all *Penicillium* species, which, in contrast to *Aspergillus* species, can grow at low temperatures (Bullerman, 1981). It is therefore of considerable importance to determine whether *Penicillium* isolates from cheese are toxigenic. This is discussed below for individual *Penicillium* mycotoxins that may be produced, such as ochratoxin A, citrinin, penicillic acid and patulin (Sections 2.2.2 and 2.2.3).

Biological testing of culture extracts has been carried out in some cases. Thus 30% of *Penicillium* isolates from Cheddar cheese (Bullerman & Olivigni, 1974) and 35% of *Penicillium* strains from Swiss cheese (Bullerman, 1976*a*) were toxic to chicken embryos. Forty-seven per cent of *Penicillium* isolates from Gouda and Cheddar cheeses in South Africa were toxigenic to ducklings (Lück *et al.*, 1976*b*). Five strains of *P. verrucosum* var. *cyclopium* from Edam cheese in Yugoslavia tested on chicken embryos were toxigenic (Škrinjar, 1985) and all of 25 extracts of mouldy Edam cheese itself (all samples contained *Penicillium*) were toxigenic in this bioassay system (Škrinjar & Žakula, 1985). Much of the toxicity in these experiments can not be explained by the presence of known mycotoxins.

In addition to those presented on Table 1 or mentioned above, studies on several other cheeses have found the presence of contaminating *Penicillium* on additional Italian cheeses, including Robbiola, Taleggio, Pannerone, Padano grano, and Provolone (Bodini *et al.*, 1969; Carini & Cerutti, 1977; Galli & Zambrini, 1978; Dragoni *et al.*, 1983) and French cheeses of various types (Dale & Guillot, 1971; Dale, 1972; Desfleurs, 1975; Lafont *et al.*, 1979*b*), as well as on powdered Dutch, Parmesan and Swiss cheeses (Northolt & de Boer, 1981), Polish cottage cheese (Burbianka & Stec, 1972), Bulgarian cheese (Dimitrov & Khadzhimitsev, 1961), Romanian cheeses (Jantea *et al.*, 1972), Russian cheese (Verdian, 1981), Czechoslovakian cheeses (Olšanský *et al.*, 1979), Iranian white cheese (Mehran *et al.*, 1975), a Canadian Cheddar cheese (Duitschaever & Irvine, 1971), and some Australian cheeses (Pitt & Hocking, 1985). *Penicillium* contaminants on *Penicillium*-ripened cheeses — Cabrales blue cheese, Gammelost (Norwegian blue cheese), Camembert and Brie — have also been documented (Keilling *et al.*, 1947, 1956; Foster *et al.*, 1957; Guegen *et al.*, 1978; Nuñez *et al.*, 1981).

Classification of penicillia has been in a continual state of change (Pitt, 1979*a*,*b*; Frisvad, 1984, 1986; Jarvis *et al.*, 1985) and in the present review, species names remain those attributed by the original investigators. Identification of *Penicillium* isolates from cheese down to species level has been carried out in many of the studies referred to above. Several of these *Penicillium* species are known as producers of mycotoxins (Scott, 1977; Leistner & Eckardt, 1979; Betina, 1984; Frisvad, 1984, 1986). For example, *P. verrucosum* var. *cyclopium* (*P. cyclopium*) can form penitrem A, α-cyclopiazonic acid, patulin, penicillic acid, xanthomegnin, viomellein and ochratoxin A; *P. crustosum* is a producer of penitrem A and roquefortine; *P. brevicompactum* is known to form brevianamide A and mycophenolic acid; and *P. viridicatum* produces numerous secondary metabolites including the mycotoxins ochratoxin A, citrinin, xanthomegnin, viomellein, viridicatumtoxin, brevianamide A, mycophenolic acid, penicillic acid, and α-cyclopiazonic acid. Mycotoxins of *P. roqueforti* have been discussed in Chapter 6.

TABLE 1
Incidence of fungal genera isolated from various cheeses

Cheese	Incidence (% of all fungal isolates)						Reference
	Penicillium	Aspergillus	Cladosporium	Alternaria	Fusarium	Other	
US cheeses	69	9	2	−[a]	—	20	Gaddi (1973)
Cheddar (USA)	82	6·6	2·0	1·1	1·1	7·0	Bullerman & Olivigni (1974)
Swiss cheese (USA)	87	0·5	—	—	—	13	Bullerman (1976a)
Mouldy cheese trimmings (USA)	93	—	4·0	—	0·9	1·8	Bullerman (1976b)
US cheeses	86	2·3	2·7	—	1·2	7·4	Bullerman (1980)
US imported cheeses	80	5·4	4·6	—	0·6	9·6	Bullerman (1980)
Mouldy surplus cheeses (USA)	100	—	—	—	—	—	Tsai & Bullerman (1985)
Home stored cheese (and other dairy products)	81	7·7	—	3·8	—	7·7	Torrey & Marth (1977a)
St Nectaire	42	—	13	3·2	—	42	Delespaul et al. (1973)

							Reference
Tome de Savoie	40	—	16	3·8	—	40	Delespaul et al. (1973)
Various Italian cheeses	78[b]	2·7	—	—	—	19	Polonelli et al. (1984)
Three Turkish cheeses	95	2·0	2·0	0·7	—	0·7	Demirer (1974)
Various cheeses (UK)	86	3·4	—	—	—	10	Chapman et al. (1983)
Various cheeses (including the above UK samples?)	89	4	2	—	—	6	Jarvis et al. (1985)
Cheese (N. Ireland)	100	—	—	—	—	—	Patterson & Damoglou (1985)
Gouda and Edam (shops, households)	84	9·4	1·6	—	0·8	4·7	Northolt et al. (1980)
Gouda and Edam (warehouses, 1976–77)	62	30	1·0	—	—	6·7	Northolt et al. (1980)
Gouda and Edam (warehouses, 1977–78)	13[c]	50[d]	(other species: 37)				Northolt et al. (1980)
Gouda and Edam (7 problem warehouses)	53 (28[e], 23[f])	34 (26[d])	—	—	—	13	Northolt & Van Egmond (1982a)
Processed cheeses (The Netherlands)	95	2·4	—	—	—	2·4	Northolt et al. (1980)

(continued)

TABLE 1—*contd.*

Cheese	Penicillium	Aspergillus	Cladosporium	Alternaria	Fusarium	Other	Reference
			Incidence (% of all fungal isolates)				
Cabrales (mould-ripened)	81	—	—	—	—	19	Nuñez et al. (1981)
Teleme (domestic Greek and imported)	83	3·8	3·3	—	1·4	8·1	Zerfiridis (1985)
White and hard cheeses (Yugoslavia)	82	2·0	7·0	2·0	—	7·0	Šutić et al. (1979)
Edam (Yugoslavia)	88	—	1·1	1·1	—	9·9	Škrinjar & Žakula (1985)
Various Egyptian cheeses	19	32	2·5	4·9	1·2	41	El-Bassiony et al. (1980)
Egyptian hard cheese	78	—	4·9	—	—	17	Abdel-Rahman & El-Bassiony (1984)
Egyptian hard cheese	26	22	2·3	—	—	50	El-Essawy et al. (1984)
Egyptian Damietta cheese	31	35	15	6·4	2·6	10	Mahmoud et al. (1983)

[a] —, not detected or not looked for.
[b] Starter cultures (*P. camemberti, P. caseicola, P. roqueforti*) excluded.
[c] *P. verrucosum* var. *verrucosum.*
[d] *A. versicolor.*
[e] *P. verrucosum* var. *cyclopium.*

2.2.2 *Penicillium* **species producing ochratoxin A and citrin**
The incidence of fungal isolates from cheese that form ochratoxin A on
further culturing in the laboratory is very low (Table 2). Most of these
isolates are (or are assumed to be) *Penicillium* species. Fungi known to
be producers of ochratoxin A include *P. verrucosum* var. *verrucosum,
P. verrucosum* var. *cyclopium* (*P. cyclopium*), *P. viridicatum,* and *P. commune*
(as well as members of the *Aspergillus ochraceus* group) (Harwig *et al.,*
1983). In addition to fungi originating from cheese (Table 2), two out of
six strains of *P. viridicatum* isolated from cheese warehouse shelves in
The Netherlands produced ochratoxin A (de Boer & Stolk-Horsthuis,
1977).

Citrinin, which often co-occurs together with ochratoxin A, was
apparently formed by isolates of *P. viridicatum* and *P. aurantiogriseum*
from cheese (Table 2) (Chapman *et al.,* 1983). However, according to
other reports, the incidence of citrinin-producing strains among
Penicillium isolates from cheese is generally low or non-existent
(Table 2).

2.2.3 *Penicillium* **species producing patulin or penicillic acid**
Patulin and penicillic acid are the most commonly produced known
mycotoxins from cheese isolates. The incidence data are presented in
Table 3. The genus of these isolates was not specified in most of the
studies by Bullerman and co-workers but are presumably *Penicillium*
spp., although the toxins can also be formed by certain *Aspergillus*
species (Scott, 1977). Even so, the incidence of fungal strains from
cheese that produce patulin and penicillic acid is not substantial so far
as research to date indicates.

2.2.4 Other toxigenic *Penicillia*
The known mycotoxins ochratoxin A, patulin and penicillic acid do not
account for toxicity of all the toxigenic *Penicillium* isolates from Swiss
and Cheddar cheeses as tested in chicken embryos (Bullerman, 1976*a*;
Bullerman & Olivigni, 1974). These studies did not include analysis for
mycophenolic acid, one of several mycotoxins toxic to the chicken
embryo (Veselý *et al.,* 1982, 1984) and which was detected in cultures of
P. brevicompactum and *P. viridicatum* isolated from French cheeses
(Lafont *et al.,* 1979*b*). Mycophenolic acid is also produced by *P.
roqueforti* and found frequently in blue-veined cheeses (Lafont *et al.,*
1979*a*) (see Chapter 6).

The hepatocarcinogenic *Penicillium* mycotoxin luteoskyrin was not

TABLE 2

Incidence of cheese isolates producing ochratoxin A and/or citrinin

Cheese	No. samples	No. or incidence[a] of toxin-producing strains		Toxigenic species/genus	Reference
		Ochratoxin A	*Citrinin*		
Cheddar (USA)	?[b]	4/287	0/287	*Penicillium* (?)	Bullerman & Olivigni (1974)
Cheddar (USA)	?	1	—[c]	*P. commune*	Pohlmeier & Bullerman (1978)
Swiss (USA)	11	0/159	0/159		Bullerman (1976a)
Mouldy cheese trimmings (USA)	?	1/304	0/304	*Penicillium*	Bullerman (1976b)
US cheeses	78	12/221	2/221	*Penicillium* (?)	Bullerman (1980)
US imported cheeses	75	10/382	2/382	*Penicillium* (?)	Bullerman (1980)

Mouldy surplus cheeses (USA)	?	Occasional	—	*Penicillium*	Tsai & Bullerman (1985)
Cheese (USA)	30	1	—	*Penicillium*	Torrey & Marth (1977*a*)
Cheesecake (Poland)	1	2	—	*P. viridicatum, Aspergillus ochraceus*	Piskorska-Pliszczyńska & Borkowska-Opacka (1984)
Edam cheese (Yugoslavia)	?	1/36	—	*P. verrucosum* var. *cyclopium*	Škrinjar (1985)
Various cheeses (UK)	52	≥10/58	≥3/58	*P. viridicatum, P. aurantiogriseum*	Chapman *et al.* (1983)
Cheese (N. Ireland)	?	0/8	0/8		Patterson & Damoglou (1985)

[a] Based on *Penicillia* isolated.
[b] ?, not stated or unclear.
[c] —, not detected or not looked for.

P. M. Scott

TABLE 3
Incidence of cheese isolates producing patulin or penicillic acid

Cheese	No. samples	No. or incidence[a] of toxigenic strains		Penicillium species	Reference
		Patulin	Penicillic acid		
Cheddar (USA)	?[b]	13/287	7/287	?	Bullerman & Olivigni (1974)
Cheddar (USA)	1	1 (both toxins)		*P. roqueforti*	Olivigni & Bullerman (1977, 1978)
Swiss (USA)	11	4/159	5/159	?	Bullerman (1976a)
Mouldy cheese trimmings (USA)	?	1/304	4/304	?	Bullerman (1976b)
US cheeses	78	5/221	15/221	?	Bullerman (1980)
US imported cheeses	75	11/382	21/382	?	Bullerman (1980)

		Main mycotoxin found	Occasional	?	
Mouldy surplus cheeses (USA)	?		Occasional	?	Tsai & Bullerman (1985)
Cheese and other dairy products (USA)	30	0/21	0/21		Torrey & Marth (1977a)
Cheese (N. Ireland)	?	0/8	0/8		Patterson & Damoglou (1985)
Various French cheeses	5/24 (patulin) 7/24 (penicillic acid)	4	6	P. expansum, P. urticae P. cyclopium	Lafont et al. (1979b)
Dutch cheeses	?	–c	0/82	P. cyclopium	Northolt et al. (1979a)

[a]Based on *Penicillia* isolated.
[b]?, not stated.
[c]–, not looked for.

detected in cheese isolates in cases where it was looked for (Bullerman, 1976*b*, 1980). The hepatotoxin rugulosin was formed by an isolate of *P. canescens* from Turkish cheese (Leistner & Pitt, 1977).

A strain of *P. crustosum* that produced the tremorgen penitrem A was isolated from mouldy cream cheese associated with intoxication of dogs (Richard & Arp, 1979). This species deserves further attention in view of its frequent occurrence on cheese and its apparent synonymity with other toxigenic *Penicillium* spp. found frequently on cheese (Pitt, 1979*a*).

2.3 *Aspergillus* Species

2.3.1 *Aspergillus versicolor*

A. versicolor and certain other *Aspergillus* species (including *A. sydowi, A. nidulans, A. flavus* and *A. ustus*) are well known as producers of sterigmatocystin, a carcinogenic mycotoxin that is structurally related to the aflatoxins (Terao, 1983). Studies describing the incidence of *A. versicolor* and/or sterigmatocystin-producing isolates from cheese are summarized in Table 4.

The occurrence of *Aspergillus* species on Edam and Gouda cheeses in Dutch warehouses has been previously referred to. During one period (1976–7), *A. repens* and *A. versicolor* were the main *Aspergillus* species isolated, although *Penicillium verrucosum* var. *cyclopium* was the most frequently isolated fungal species overall (Northolt *et al.*, 1980). In 1977–8, however, *A. versicolor* was the most frequently isolated fungal species, a significant finding. *A. versicolor* (and *A. repens*) were more associated with riper cheeses during both periods. It can be concluded that the strains of *A. versicolor* isolated were sterigmatocystin producers, either because the toxin was present in the cheese or the isolate formed sterigmatocystin on malt extract agar (Northolt *et al.*, 1980). Subsequently, Northolt & Van Egmond (1982*a*) found that cheeses from seven warehouses previously shown to harbour *A. versicolor* were again frequently contaminated with *A. versicolor. A. versicolor* was not isolated from mouldy cheeses in factories where no plastic coating was used on the cheese (Northolt & Van Egmond, 1982*a*). Additionally, most samples taken from air, cheese shelves and transport carts in warehouses in The Netherlands contained *A. versicolor* (Northolt & Van Egmond, 1982*a*), and five strains of *A. versicolor* that produced sterigmatocystin were isolated from cheese shelves (de Boer & Stolk-Horsthuis, 1977). However, apart from this and the other Dutch studies, the evidence is that sterigmatocystin-producing fungal isolates are infrequent cheese contaminants in general (Table 4).

In addition to sterigmatocystin, *A. versicolor* is known to be a producer of the mycotoxins α-cyclopiazonic acid and versicolorin A (Hendricks *et al.*, 1980; Cole & Cox, 1981).

2.3.2 *Aspergillus flavus*

Although of minor importance from the point of view of incidence and growth on cheese, *A. flavus* is of great concern because it produces the highly hepatocarcinogenic aflatoxins; the other main aflatoxin-producing fungus, *A. parasiticus*, only appears to have been isolated from cheese in two instances (Verdian, 1981). *A. flavus* is also a known source of several other mycotoxins, including α-cyclopiazonic acid, sterigmatocystin, aflatrem, aspergillic acid, kojic acid, and β-nitropropionic acid (Cole & Cox, 1981). Instances of occurrence of *A. flavus* on cheese and whether the strains were aflatoxigenic are summarized in Table 5. These probably do not represent complete information on prevalence of *A. flavus* in cheese as many studies (Table 1) only identified the genera of isolated fungi, although aflatoxin production was determined in some cases (Bullerman, 1981). It can certainly be concluded that *A. flavus* is not a common species on cheese. One possible special source of inoculum, if not sterilized, is pepper added to certain cheeses (Jacquet & Teherani, 1974).

2.3.3 Other toxigenic *Aspergilli*

Although the incidence of *Aspergillus* strains isolated from Cheddar cheese was only 6·6% of all fungal isolates, 48% of them were toxic to chicken embryos (Bullerman & Olivigni, 1974), indicating formation of mycotoxins. Important *Aspergillus* species other than *A. versicolor* and *A. flavus* that have been isolated from a wide variety of cheeses from different parts of the world include *A. ustus*, members of the *A. glaucus* group (*A. manginii, A. amstelodami* and the teleomorph of *A. ruber, Eurotium ruber*), *A. candidus, A. niger*, and *A. fumigatus* (Boutibonnes & Jacquet, 1969; Jacquet *et al.*, 1970; Burbianka & Stec, 1972; Demirer, 1974; Mehran *et al.*, 1975; Lück *et al.*, 1976*b*; Domenichini, 1978; Galli & Zambrini, 1978; El-Bassiony *et al.*, 1980; Northolt *et al.*, 1980; Verdian, 1981; Mahmoud *et al.*, 1983; Naguib *et al.*, 1983; Polonelli *et al.*, 1984). The high frequency of isolation of some of these species on certain Egyptian cheeses (El-Bassiony *et al.*, 1980; Mahmoud *et al.*, 1983) is noteworthy in view of the generally low incidence of *Aspergilli* on cheese

P. M. Scott

TABLE 4
Incidence of *Aspergillus versicolor* and producers of sterigmatocystin on cheese

Cheese	No. samples	No. or incidence[a] of strains isolated		Reference
		A. versicolor	Sterigmatocystin producers	
Gouda/Edam (shops, households, The Netherlands)	101	1/128	1	Northolt *et al.* (1980)
Gouda/Edam (warehouses, 1976–77)	114	24/208	24	Northolt *et al.* (1980)
Gouda/Edam (warehouses, 1977–78)	140	75/150	75	Northolt *et al.* (1980)
Gouda/Edam (7 problem warehouses)	67	42	—[b]	Northolt & Van Egmond (1982a)
Processed cheese (shops, The Netherlands)	39	0	—	Northolt *et al.* (1980)

		2 (+1 *A. sydowi*)	3	
Various cheeses (France)	2/24			Lafont *et al.* (1979*b*)
Provolone	?[c]	+[d]	–	Galli & Zambrini (1978)
Fiore Sardo	?	+	–	Domenichini (1978)
Three Turkish cheeses	91	2/148	–	Demirer (1974)
Cheddar (USA)	?	–	0/349	Bullerman & Olivigni (1974)
Swiss (USA)	11	–	0/183	Bullerman (1976*a*)
Domestic and imported (USA)	153	–	0/735	Bullerman (1980)
Various cheeses (USA)	49	0	–	Stoloff (1982)
Damietta cheese (Egypt)	40	8·5%	–	Mahmoud *et al.* (1983)
Blue cheese (Egypt)	5	1	1	Moubasher *et al.* (1978)

[a] Based on all strains of fungi isolated.
[b] –, not looked for.
[c] ?, not stated.
[d] +, present but incidence not quantitated.

TABLE 5
Incidence of *Aspergillus flavus* and aflatoxin-producing fungi on cheese

Cheese	No. samples	No. or incidence[a] of strains		Reference
		A. flavus	Aflatoxin producers	
Various (Canada)	24	1 (*A. flavus* var. *columnaris*)	1 (aflatoxin B_2)	Van Walbeek *et al.* (1968)
Cheese and other dairy products	30	$-$[b]	2/26	Torrey & Marth (1977*a*)
Cheddar	?[c]	$-$	1/349	Bullerman & Olivigni (1974)
Swiss cheese	11	1/183	1 (B_1, B_2)	Bullerman (1976*a*)
US cheeses	78	2/256	2 (B_1, B_2, G_1, G_2)	Bullerman (1980)
US imported cheeses	75	11/479	11 (B_1, B_2, G_1, G_2)	Bullerman (1980)
Gouda/Edam	355	0/486	$-$	Northolt *et al.* (1980)
Gouda/Edam	67	0	$-$	Northolt & Van Egmond (1982*a*)
Romanian cheeses	109	38	0/38	Mihai *et al.* (1970)
Teleme	94	$-$	0/211	Zerfiridis (1985)
Cheese	1	1	1 (B_1)	Vanderhoven *et al.* (1970)

Pepper cheese	1	1	1	Jacquet & Teherani (1974)
Gruyère	12	2	0	Jacquet *et al.* (1970)
St. Nectaire	?	Occasional	—	Dale & Guillot (1971)
Cheesecake (Poland)	1	+[d]	—	Piskorska-Pliszyńska & Borkowska-Opacka (1984)
Cottage cheese (Poland)	43	+	—	Burbianka & Stec (1972)
Kachkaval cheese	?	1	—	Dimitrov & Khadzhimitsev (1961)
Kachkaval cheese	?	+	—	Verdian (1981)
Poshekhonsky cheese	?	+	—	Verdian (1981)
Cheese (Yugoslavia)	?	1/100	1	Šutić *et al.* (1979)
Egyptian Roquefort cheese	62	260	39 (B_1, B_2)	Naguib *et al.* (1983)
Egyptian Damietta cheese	40	≥4	—	Mahmoud *et al.* (1983)
Egyptian cheese	1	2	2 (B_2)	El-Bazza *et al.* (1983)
Egyptian cheeses	84	6	1	El-Bassiony *et al.* (1980)

[a] Based on all strains of fungi isolated.
[b] —, not looked for.
[c] ?, not stated.
[d] +, present but incidence not quantitated.

(Table 1). Mycotoxins known to be produced by the above species include austdiol (*A. ustus*) (Lück & Wehner, 1979); physcion and erythroglaucin (*A. glaucus* group) (Bachman *et al.*, 1979; Podojil *et al.*, 1979); AcT₁, kojic acid, citrinin and xanthoascin (*A. candidus*) (Takahashi *et al.*, 1976; Cole & Cox, 1981; Chattopadhyay *et al.*, 1987); malformins A₁ and C, nigragillin, aurasperone D and oxalic acid (*A. niger*) (Ghosal *et al.*, 1979; Cole & Cox, 1981); and fumigatoxin, gliotoxin, and various tremorgenic mycotoxins (*A. fumigatus*) (Iwata *et al.*, 1969; Cole & Cox, 1981). General reviews on mycotoxins from *Aspergillus* spp. have been published (Moss, 1977; Mislivec, 1985).

2.4 Other Fungal Genera Isolated from Cheese

The occasional presence of fungi other than *Penicillium* and *Aspergillus* on cheese is indicated in Table 1, which includes some incidence data on *Cladosporium, Alternaria,* and *Fusarium.*

Incidences of *Cladosporium* as high as 13–16% of the normal fungal population on some St. Nectaire and Tome de Savoie cheeses have been reported (Delespaul *et al.*, 1973) (Table 1); these were mainly *C. cladosporioides* and *C. herbarum*. Additional to those listed on Table 1, isolations of *Cladosporium* from cheese have been recorded by Jacquet & Desfleurs (1966*b*), Dale & Guillot (1971), Dale (1972), Moreau (1976*b*), Moubasher *et al.* (1978), Verdian (1981) and Jarvis *et al.* (1985). Moubasher *et al.* (1978) found that their isolates (*C. cladosporioides*) were toxigenic by the chicken embryo test. Known mycotoxins from *Cladosporium* species include epicladosporic acid and fagicladosporic acid (Bilai, 1960) and emodin (Gross *et al.*, 1984). Allergens have also been isolated from *Cladosporium* species (Landmark & Aukrust, 1985; Vijay *et al.*, 1985).

Alternaria species are an important source of mycotoxins (King & Schade, 1984) but occurrence on cheese is rare (Table 1; Dale, 1972; Moubasher *et al.*, 1978). *Fusarium* species are also infrequent on cheese (Table 1; Moreau, 1976*b*) — this genus is well known for production of many mycotoxins such as trichothecenes, zearalenone and moniliformin (Cole & Cox, 1981; Ueno, 1985).

Geotrichum candidum (*Oidium lactis, Oospora lactis*) is a mould that occurs on numerous soft cheeses such as Camembert, Limbourg, Robbiola, and Taleggio (Foster *et al.*, 1957; Carini & Cerutti, 1977; Polonelli *et al.*, 1984) as well as on cottage cheese (Foster *et al.*, 1957). *G. candidum* was the dominant mould contaminant of some

Czechoslovakian cheese and other dairy products (Olšanský *et al.*, 1979). It is also found on blue cheese (Nuñez *et al.*, 1981) and on the surface of hard cheeses such as Saint-Nectaire, forming a sought-after grey film and preventing contamination by other fungi (Dale & Guillot, 1971; Dale, 1972; Delespaul *et al.*, 1973). *Geotrichum* was one of several moulds isolated from Egyptian cheeses (Mahmoud *et al.*, 1983; El-Essawy *et al.*, 1984). Although *G. candidum* is not generally regarded as a mycotoxin producer, ergot alkaloids have been formed by this species (El-Refai *et al.*, 1970).

Fungi of the genus *Mucor* have been reported to occur on various kinds of cheese (Foster *et al.*, 1957; Augusto *et al.*, 1968; Bodini *et al.*, 1969; Dale & Guillot, 1971; Brenet *et al.*, 1972; Dale, 1972; Delespaul *et al.*, 1973; Gaddi, 1973; Moreau, 1976*b*; Carini & Cerutti, 1977; Torrey & Marth, 1977*a*; Ölsanský *et al.*, 1979; Šutić *et al.*, 1979; El-Bassiony *et al.*, 1980; Verdian, 1981; Dragoni *et al.*, 1983; Mahmoud *et al.*, 1983; Naguib *et al.*, 1983; Abdel-Rahman & El-Bassiony, 1984; El-Essawy *et al.*, 1984; Polonelli *et al.*, 1984; Škrinjar, 1985; Škrinjar & Žakula, 1985; Zerfiridis, 1985). *Mucor* can rapidly invade cheese and has been known to cause defects ('Poil de Chat' or 'Cat's Fur') in Camembert cheese (Brenet *et al.*, 1972; Moreau, 1976*b*; Recordon *et al.*, 1980), probably due to poor acidification (Sozzi & Shepherd, 1971). Certain species of *Mucor* and *Rhizopus*, another member of the Mucorales which occasionally contaminates cheese (Moubasher *et al.*, 1978; Northolt *et al.*, 1980; Mahmoud *et al.*, 1983; El-Essawy *et al.*, 1984; Jarvis *et al.*, 1985), are toxigenic (Wilson *et al.*, 1984; Rabie, 1986). Rhizonin A is the major mycotoxin known to be formed by *Rhizopus* (*R. microsporus*) (Wilson *et al.*, 1984); ergot alkaloids have been detected in *R. nigricans* and *M. hiemalis* (El-Refai *et al.*, 1970).

Other potentially mycotoxigenic fungi that have been isolated from cheese of various kinds include *Drechslera* (Moubasher *et al.*, 1978), *Oospora* (Burbianka & Stec, 1972; Verdian, 1981), *Acremonium* (*Cephalosporium*) (Gaddi, 1973; El-Bassiony *et al.*, 1980; El-Essawy *et al.*, 1984; Škrinjar & Žakula, 1985), *Trichoderma* (Ölsanský *et al.*, 1979; Verdian, 1981); *Rhizoctonia* (Dale, 1972), *Cylindrocarpon* (Delespaul *et al.*, 1973), *Verticillium* (Pitt & Hocking, 1985), *Gliocladium* (Naguib *et al.*, 1983) and *Scopulariopsis* (Jacquet & Desfleurs, 1966*a*; Gaddi, 1973; Galli & Zambrini, 1978; Northolt *et al.*, 1980; Northolt & Van Egmond, 1982*a*; Chapman *et al.*, 1983; Dragoni *et al.*, 1983; El-Essawy *et al.*, 1984). *Drechslera* (formerly known as *Helminthosporium*) produces several mycotoxins, including sterigmatocystin (Shotwell & Ellis, 1976;

Schneider *et al.*, 1985); *Oospora* and *Acremonium* form oosporein (Cole & Cox, 1981); slaframine and swainsonine are *Rhizoctonia* toxins (Broquist, 1986); *Verticillium* produces verticillin A (Cole & Cox, 1981); and *Gliocladium* and *Trichoderma* produce gliotoxin (Cole & Cox, 1981). *Cephalosporium, Trichoderma* and *Cylindrocarpon* species are known trichothecene producers (Ichinoe & Kurata, 1983) while *Scopulariopsis* is also considered as toxigenic (Dragoni *et al.*, 1983).

2.5 Conclusions

Incidence studies on moulds in a wide variety of cheeses indicate that *Penicillium* species are the most commonly found, particularly on refrigerated cheese. Mycotoxins that have been detected in *Penicillium* species isolated as contaminants from cheese are ochratoxin A, citrinin, patulin, penicillic acid, mycophenolic acid and penitrem A, although the incidence of toxigenic strains is usually very low. Biological testing of *Penicillium* isolates, however, has indicated a much higher incidence of toxigenic isolates. *Aspergillus* species are generally recovered in low incidence from cheese but exceptional studies on Dutch, Egyptian and Italian cheese have been reported. Mycotoxins known to be formed by *Aspergillus* isolates from cheese are sterigmatocystin and aflatoxins; *A. flavus* is not common on cheese, however. Other important fungi that occur on cheese include *Cladosporium, Geotrichum* and *Mucor*. More work by chemists and toxicologists is required to follow up on mycological work on cheeses, particularly those with naturally mouldy rinds.

3 ABILITY OF FUNGI TO PRODUCE MYCOTOXINS ON CHEESE

3.1 General

Cheese is an adequate substrate for mould growth given suitable conditions of temperature and moisture. However, little comparative work has been done on the relative ability of different types of cheese to support actual fungal growth. From studies with *Aspergillus parasiticus* and *Penicillium camemberti* (*caseicolum*), aged Cheddar was the most inhibitory to mould growth of four cheese types tested (Yousef & Marth, 1987). Mycotoxin-producing moulds require oxygen (Bullerman *et al.*, 1984) and thus packaging of cheese is an important factor. Stott &

Bullerman (1976) found that Cheddar cheese inoculated with *P. patulum* did not support fungal growth when vacuum packaged or coated with paraffin wax but growth was heavy when the cheese was packaged in air. Having said this, it should be noted that the experiments reviewed in this section were carried out under aerobic conditions. Also it must be emphasized that the presence of fungal growth does not necessarily imply concomitant formation of mycotoxins.

Concentrations of various mycotoxins formed in cheese are lower than in yeast extract–sucrose medium (Engel, 1978). Although most work on the topic of mycotoxin production on cheese has been carried out with aflatoxin-producing *Aspergilli*, the formation of *Penicillium* mycotoxins experimentally on cheese will be reviewed first in accordance with the importance of this genus as a cheese contaminant.

3.2 *Penicillium* Toxins

3.2.1 Ochratoxin A and citrinin
There is some experimental evidence that *Penicillium* spp. can produce ochratoxin A and citrinin on cheese. On Edam cheese slices of water activity (a_w) 0·95, *P. cyclopium* (*P. verrucosum* var. *cyclopium*) grew at temperatures down to 0°C but formed ochratoxin A (up to 0·024 µg/g) only at 20°C and 24°C (Northolt *et al.*, 1979*b*; Northolt & Van Egmond, 1982*b*). In another study on Edam cheese, however, an ochratoxigenic strain of *P. verrucosum* var. *cyclopium* failed to produce ochratoxin A on the cheese at temperatures of 26–28°C (one week) followed by 10°C (4–6 weeks) (Škrinjar, 1985). A strain of *P. commune* produced up to 0·6 µg/g of ochratoxin A on Cheddar cheese incubated at 25°C and lower concentrations at 12°C and 5°C (Pohlmeier & Bullerman, 1978; Bullerman, 1981). Details of these experiments have not been published, however. A strain of *P. verrucosum* var. *verrucosum* produced up to 35 µg/g of ochratoxin A on Gouda (but not Edam) cheese at 25°C (Nowotny *et al.*, 1983*b*); however, another strain of the same species that produced both ochratoxin A and citrinin on maize, did not produce these toxins on cheese (Nowotny *et al.*, 1983*b*). Formation of ochratoxin A (0·1 µg/g) on cheese by *P. verrucosum* var. *verrucosum* was also noted by Yazaki *et al.* (1980).

Occurrence of ochratoxin A (up to 0·28 µg/g) and citrinin (up to 0·25 µg/g) in unsterilized Cheddar cheese blocks after inoculation with *P. aurantiogriseum* or *P. viridicatum* and incubation for 10 days at 25°C was probably due to these fungi (Chapman *et al.*, 1983; Jarvis, 1983), although other moulds grew during the experiments. Considerable variation in mycotoxin levels was observed among 10 inoculated cheese samples. Levels of citrinin as high as 450 μg/g on Gouda cheese and up to about 80 μg/g on Edam cheese were formed by a strain of *P. citrinum* at 25°C (Nowotny *et al.*, 1983*b*). However, Engel (1978) found that a strongly toxigenic strain of *P. citrinum* did not produce citrinin in autoclaved Tilsit cheese at 27°C; *A. ochraceus* formed 1·2 μg/g of ochratoxin A on this cheese. There is a need for more work on the production of ochratoxin A (and citrinin) by *Penicillium* spp. growing on cheese at refrigerator temperatures.

3.2.2 Patulin and penicillic acid

It is apparent that patulin and penicillic acid can occur naturally in cheese (see Section 5.2) in spite of their instability in this substrate (see Section 4.4). Results of inoculation experiments appear to be contradictory, however. A patulin-producing strain of *P. patulum* grew well on Cheddar cheese at about 5°C and 25°C but patulin was only detected at 25°C, in low and variable concentrations (Stott & Bullerman, 1976). Penicillic acid was detected in small amounts in slices of Gouda cheese (a_w 0·98), but not Tilsit cheese (a_w 0·96), incubated for six weeks at 16°C with *P. cyclopium* (Northolt *et al.*, 1979*a*). All other experiments that have been carried out to determine whether patulin or penicillic acid form in cheese have been negative, in spite of good mould growth (Lieu & Bullerman, 1977; Olivigni & Bullerman, 1977; Engel, 1978; Engel & Prokopek, 1980). Cheese substrates used were Tilsit (autoclaved), blue cheese, Swiss cheese, Mozzarella, and Cheddar cheese while *Penicillium* species used for inoculation were *P. urticae, P. patulum, P. claviforme*, and *P. roqueforti* for patulin and *P. viridicatum* and *P. roqueforti* for penicillic acid. Incubation temperatures ranged from 5–27°C and times up to 80 days. The main reasons suggested for little or no production of patulin and penicillic acid in cheese were suboptimal a_w values (Northolt *et al.*, 1978, 1979*a*; Northolt & Van Egmond, 1982*b*) and unsuitability of a substrate low in carbohydrate and high in protein (Olivigni & Bullerman, 1977).

3.2.3 Other *Penicillium* mycotoxins

Penicillium puberulum has been shown to form aflatoxin B$_1$ in Tilsit cheese (Kiermeier & Groll, 1970*b*). Fortunately, production of aflatoxin by *Penicillium* species appears to be extremely rare and its identification in this genus is in fact debatable (Mislivec *et al.*, 1968). In this case, aflatoxin was formed after 18 days at only 5°C at a relative humidity of 99%; more was present at 20°C.

According to the work of Engel (1978), who studied a variety of mycotoxigenic fungal species (both *Aspergillus* and *Penicillium*), the mycotoxin formed in the highest concentrations in autoclaved Tilsit cheese was citreoviridin. *P. citreoviride* produced it at 27°C (30 μg/g) and 16°C (10 μg/g) after 20 days' incubation.

Growth of *P. crustosum* on cream cheese at 27°C for two weeks, followed by one week at 4°C yielded penitrem A (Richard & Arp, 1979).

3.3 *Aspergillus* Toxins

3.3.1 Sterigmatocystin
Infection of the surface of Wilstermarsch and Edam cheeses with two strains of *Aspergillus nidulans* (a species that does not appear to have been recorded on cheese) resulted in production of sterigmatocystin (up to 1·75 μg/g in the first 1-mm layer) only in the Wilstermarsch cheese at room temperature (Engel & Teuber, 1980). There was mould growth at 15°C but no sterigmatocystin was detected. When two strains each of *A. versicolor* and *A. nidulans* were grown on autoclaved Tilsit cheese, up to 0·19 μg/g of sterigmatocystin was formed at 27°C by *A. nidulans* but no significant quantities were produced by *A. versicolor* (Engel, 1978; Engel & Teuber, 1980), although all four strains were mycotoxigenic in liquid medium (Engel & Teuber, 1980). One strain of *A. nidulans* also yielded sterigmatocystin at 15°C. Similar observations, that *A. versicolor* grew well on Edam and Gouda cheeses at 25°C but did not produce sterigmatocystin, were made by Nowotny *et al.* (1983*b*). These results with *A. versicolor* are in contrast to findings of Gouda and Edam cheeses being naturally infected in warehouses with *A. versicolor* and contaminated with sterigmatocystin (Northolt *et al.*, 1980; Northolt & Van Egmond 1982*a*). Growth of *A. versicolor* and formation of sterigmatocystin (up to 0·6 μg/g in the outer 1-cm layer) occurred at various warehouse temperatures; an a_w range of 0·84–0·93 for toxin production was observed (Northolt & Van Egmond, 1982*a*).

3.3.2 Aflatoxins B$_1$, B$_2$, G$_1$ and G$_2$
In spite of the extremely rare occurrence of aflatoxigenic *Aspergillus* spp. on cheese, considerable effort has gone into determining experimental conditions under which *A. flavus* or *A. parasiticus* produce aflatoxins on

cheese. Aflatoxin production on cheese has been reviewed in detail by Kiermeier (1971*a*, 1981), Bullerman (1981), Applebaum *et al.* (1982) and Tantaoui-Elaraki & Khabbazi (1984). See also Section 3.2.3. The following cheeses have been shown to be substrates for aflatoxin production by *A. flavus* or *A. parasiticus*: Tilsit (Frank, 1966, 1968; Kiermeier & Groll, 1970*b*; Rothenbühler & Bachman, 1970; Kiermeier & Behringer, 1977*b*; Kiermeier & Rumpf, 1975; Engel, 1978); Emmenthal (Rothenbühler & Bachman, 1970; Kiermeier & Rumpf, 1975; Jacquet & Tantaoui-Elaraki, 1976); Cheddar (Lie & Marth, 1967; Oldham *et al.*, 1971; Park & Bullerman, 1983); Swiss cheese (Lieu & Bullerman, 1977); Mozzarella (Lieu & Bullerman, 1977); brick cheese (Shih & Marth, 1972*a*); Teleme cheese (Zerfiridis, 1985); Provolone (Kiermeier & Behringer, 1977*b*); white brined Feta cheese (Karaioannoglou, 1984); 'Manchego' cheese (Pérez *et al.*, 1973); Gouda (Jacquet & Tantaoui-Elaraki, 1976); various French cheeses, including Boursin (garlic and fine herbs), and fromage des Pyrenées (Jacquet & Tantaoui-Elaraki, 1976); and Romanian pressed cheese (Mihai *et al.*, 1970). Additional cheese substrates that support aflatoxin production are processed cheese (Frank, 1966; Polzhofer, 1977*a*; Park & Bullerman, 1983), cottage cheese (Burzyńska, 1977; Park & Bullerman, 1983), powdered cheese (Kiermeier & Rumpf, 1975), and a cheese powder paste (Kiermeier & Behringer, 1977*a*).

Although casein stimulates aflatoxin production by *A. parasiticus* in liquid medium (Kiermeier & Zierer, 1975), in general cheese is a fairly poor substrate for aflatoxin formation; Wildman *et al.* (1967) noted that it ranked below many other foods. However, high yields of up to 318 $\mu g/g$ total aflatoxins have been obtained on Cheddar cheese (Park & Bullerman, 1983) and 331 $\mu g/g$ on goat's cheese without crust (Jacquet & Tantaoui-Elaraki, 1976).

In cheese that already contained internal or external mould, e.g. blue cheese, Brie, Camembert, French goat cheese, or Pont-L'Évêque, *A. flavus* did not grow and aflatoxins were not detected (Kiermeier & Groll, 1970*b*; Jacquet & Tantaoui-Elaraki, 1976). If the external mould was removed from the cheese then growth of *A. flavus* and aflatoxin production occurred on the remaining cheese, with the exception of Brie and Maroilles (where there was still no growth of *A. flavus*) and Munster and Reblochon, where there was some fungal growth but no aflatoxin production (Jacquet & Tantaoui-Elaraki, 1976). This effect was quite striking for French goat cheese: the whole cheese did not support growth of *A. flavus* but following removal of its exterior crust before inoculation, the cheese contained 331 µg/g of aflatoxins after incubation for seven days at ambient temperature. Growth of *A. flavus*, but no aflatoxin production, was observed on Edam, Boursin (with pepper), Mimolette, and Reblochon (without rind) (Jacquet & Tantaoui-Elaraki, 1976). No *A. flavus* growth or aflatoxin production occurred on Romadur cheese, which was attributed to smear formation (Kiermeier & Groll, 1970*b*).

The lower limits of temperature for growth of *A. flavus* or *A. parasiticus* and for aflatoxin production are important practical factors in cheese storage and the values found for different strains on various cheeses after a sufficient incubation time at adequate moisture levels are given in Table 6. Most studies showed that aflatoxins can only be produced on

TABLE 6
Lower limits of temperature for growth of *Aspergillus flavus* and *A. parasiticus* and production of aflatoxin on cheese

Aspergillus *sp.*	*Cheese*	Temperature limit (°C)		*Reference*
		Growth	*Aflatoxin production*	
A. flavus, *A. parasiticus*	Tilsit	16	18	Kiermeier & Groll (1970*b*)
A. flavus, *A. parasiticus*	Tilsit	16	16	Engel (1978)
A. flavus	Swiss	>12	>12	Lieu & Bullerman (1977)
A. flavus	Cheddar	>7	>7	Oldham *et al.* (1971)
A. flavus, *A. parasiticus*	Cheddar	−[a]	15	Park & Bullerman (1983)
A. flavus	Brick	7·2	>13<24	Shih & Marth (1972*a*)
A. parasiticus	Brick	7·2	>7·2<13	Shih & Marth (1972*a*)
A. flavus	Feta	>10<13	>10<13	Karaiannoglou (1984)
A. parasiticus	Cottage cheese	−	>15	Park & Bullerman (1983)
A. flavus	Cottage cheese	−	15	Park & Bullerman (1983)

[a] −, not reported.

cheese at temperatures above 10°C. In practice, ripening temperatures for cheeses will often be >10°C. The optimum temperature for aflatoxin B_1 formation on Tilsit cheese was found to be 26°C (Kiermeier & Groll, 1970b). Mean air temperatures of 3·9–11·9°C have been recorded in household refrigerators (Torrey & Marth, 1977b) and aflatoxin formation in refrigerated cheese is unlikely to occur. Some doubt is thrown on observations by Kiermeier & Behringer (1977a) of low level aflatoxin production at 5°C on a Tilsit cheese powder paste, because of possible carryover of toxin from the spore inoculum.

A limiting a_w of 0·79 was found for growth of *A. flavus* and aflatoxin production on Tilsit cheese at 25°C (Kiermeier & Groll, 1970b); production rose sharply at higher relative humidities. Grated cheeses, which are often stored at room temperature, do not normally contain sufficient moisture to support mould growth (Lie & Marth, 1967).

Processed cheese was a good substrate for aflatoxin production by *A. flavus* as shown by Polzhofer (1977a) but not according to Park and Bullerman (1983). The difference may be due to emulsifying salts or sodium chloride, which at concentrations of >3% caused marked decreases in aflatoxin concentrations in cultures on processed cheese (25°C) (Polzhofer, 1977a) and in liquid medium (at 21°C) (Shih & Marth, 1972b). Sodium chloride (1–2%) the salt concentration of most cheeses, slightly increased aflatoxin yields in the liquid medium (Shih & Marth, 1972b).

Studies with inoculated cottage cheese curd from skim milk at various initial pH values showed that the highest amounts of aflatoxins B_1 and G_1 were formed after 21 days at extreme initial pH values of 1·7–2·4 and 8·3–9·9, although growth of *A. flavus* and *A. parasiticus* was highest at intermediate initial pH values (Lie & Marth, 1968).

3.3.3 β-Nitropropionic acid
β-Nitropropionic acid has been produced on cheese curds inoculated with *Aspergillus oryzae* and incubated at 28°C; maximum concentrations found were 203–427 μg/g after five days (Iwasaki & Kosikowski, 1973).

3.4 Prevention of Fungal Growth

Mould growth on cheese can be prevented or delayed by addition of antifungal agents such as sorbic acid (or sorbate), propionic acid, natamycin (pimaricin), and mycostatin (Wallhäuser & Lück, 1970; Nilson *et al.*, 1975; Bullerman, 1977*a*; Mattsson, 1977; Galli *et al.*, 1978; Lück & Cheesman, 1978; Boersma, 1979; Morris & Castberg, 1980; Andres, 1982; Ray & Bullerman, 1982; Yousef & Marth, 1987). Several years of continuous use of natamycin in cheese warehouses did not desensitize the moulds to this antifungal agent (de Boer & Stolk-Horsthuis, 1977). Comparative studies showed that potassium sorbate was more effective than pimaricin and propionic acid on Provolone cheese (Galli *et al.*, 1978); pimaricin was slightly more effective than mycostatin in cottage cheese (Nilson *et al.*, 1975); and pimaricin was more active than sorbate in laboratory medium (Bullerman, 1977*a*). Pimaricin inhibited different strains of *A. flavus* and *A. parasiticus* to various degrees (Kiermeier & Zierer, 1975) but cheeses themselves would have to be treated early, before mould growth could start, as pimaricin does not stop aflatoxin production *per se*. Also, pimaricin requires repeated application on ripening cheeses to be effective. A Swiss commercial mixture ('Sezet') consisting of a complex plant extract in a polymer dispersion prevented mould growth on several Swiss cheeses (Jully, 1977). Of course, good plant sanitation during cheese manufacture and handling is a better way of minimizing or preventing mould growth on cheese (Bullerman, 1977*b*).

3.5 Conclusions

Cheese is generally a good substrate for fungal growth but a poor substrate for experimental mycotoxin production. There is some evidence that the *Penicillium* mycotoxins ochratoxin A, citrinin, patulin, penicillic acid, citreoviridin, and penitrem A can be formed on cheese but the experiments are sometimes contradictory. Sterigmatocystin has been produced on cheese by *Aspergillus nidulans*, but no significant quantities have been formed experimentally by *A. versicolor* on cheese, contrary to findings under natural storage. Other *Aspergillus* mycotoxins that have been produced on cheese are the aflatoxins and β-nitropropionic acid. The aflatoxins can be formed on many different types of cheese at >7°C, usually in low concentrations, although up to 331 µg/g has been recorded. Mould growth on cheese can be prevented or delayed by early treatment with antifungal agents such as pimaricin (natamycin) or sorbate.

4 PENETRATION AND STABILITY OF MYCOTOXINS IN CHEESE

4.1 General

Faced with a block or piece of cheese on which surface mould has developed, the consumer or manufacturer must make a risk/benefit decision whether to trim the cheese or throw it away. Usually it is trimmed, although Bullerman (1981) advised against this if the cheese had been improperly stored at high temperatures, with a possibility of aflatoxin contamination. The colour of the mould — olive green, dark forest green or dull green may indicate the potential aflatoxin or sterigmatocystin producers *Aspergillus flavus, A. parasiticus* or *A. versicolor* may be present. As well, the extent of mould growth, cheese size, and type of cheese are additional factors to be considered in the decision (Bullerman, 1981). These factors also should be taken into account if deciding how much cheese to trim off. This question can not be left in the hands of the producer because his decision may be economically influenced. Neither can it always be in the hands of the consumer who has no knowledge of the history of the cheese or any possible dangers. The problem should be managed by health authorities who would make informed decisions for inspection and control. The required scientific knowledge includes that on migration of mycotoxins in cheese. A number of experiments have been carried out to determine how far mycotoxins penetrate into cheese. These are mainly studies using cheeses inoculated with toxigenic fungi and have usually concerned the aflatoxins, although some experiments have involved sterigmatocystin, ochratoxin A, citrinin, patulin and penicillic acid.

4.2 Migration of Mycotoxins in Cheese

4.2.1 Ochratoxin A and citrinin

Bullerman (1981) reported that ochratoxin A penetrated to a depth of 0·7 cm in Cheddar cheese inoculated with *Penicillium commune* (Pohlmeier & Bullerman, 1978) and incubated at 25°C. Migration studies on a large block of cheese (dimensions not given) naturally contaminated with ochratoxin A and citrinin indicated that ochratoxin A penetrated 2 cm (surface and core) and even 8 cm (surface only, traces) from the end of the cheese with active mould growth; traces of citrinin were also present up to a distance of 8 cm (surface and core) (Cooper *et al.*, 1982). In a more detailed study in the same laboratory, strains of *P. aurantiogriseum* (*P. cyclopium*) or *P. viridicatum* were inoculated onto two opposite faces of ten 500 g blocks of Cheddar cheese and incubated at 25°C for 10 days (Chapman *et al.*, 1983; Jarvis, 1983). Results showed that 10/10 samples taken from the cheese core contained ochratoxin A at levels of 20–280 ng/g compared to 9/10 samples taken from the 0·5 cm surface layer. The least distance from the surface of the blocks of cheese to the core material was about 2 cm. There were more instances where the level of ochratoxin A in the core samples exceeded (or equalled) the level in the surface layer than the reverse (Chapman *et al.*, 1983). In the case of citrinin, it was found in 3/10 core samples compared to 1/10 surface samples (at levels of 100–250 ng/g). These findings must be reviewed with some caution as the uninoculated cheese contained about 20 ng/g of ochratoxin A and possibly some citrinin; mould growth also occurred on incubation of this control cheese with an increase in ochratoxin A concentration to 70 ng/g in the core material. Nevertheless it appears that migration of ochratoxin A and citrinin into cheese can occur to a considerable extent at 25°C. It should be noted that no migration experiments with these mycotoxins at refrigerator temperatures have been reported in the literature.

4.2.2 Patulin and penicillic acid

While no patulin was detected in the centre of a sample of hard cheese (dimensions not known) that was naturally contaminated in the outer 2 cm layer with 90 ng/g of patulin, penicillic acid was measurable (45 ng/g) in the centre of 1/4 naturally-contaminated hard cheeses (Lafont *et al.*, 1979*b*); this sample contained 340 ng/g of penicillic acid in the outer 2 cm zone.

A sample obtained from the first 3 mm of a Cheddar cheese block moulded at 25°C with *Penicillium patulum* contained 200 ng/g of patulin, but none was detected in the mycelium or in slices deeper than 3 mm into the cheese (Stott & Bullerman, 1976). The instability of patulin in cheese (see Section 4.4) would markedly affect any migration in cheese.

4.2.3 Sterigmatocystin

The migration of sterigmatocystin in Wilstermarsch cheese after inoculation with *Aspergillus nidulans* and incubation at 15°C or room temperature for 40 days was limited to 0·5 cm from the cheese surface (Engel & Teuber, 1980). Similar results were obtained with six naturally-contaminated Gouda cheeses on which *A. versicolor* was growing (Van

Egmond *et al.*, 1982*b*). Cheeses varied in age from 2–8 months. In only one case, where the surface scrapings (0·1–0·5 mm) contained 230 ng sterigmatocystin/cm^2 and the rind (*c.* 2 mm) 800 ng/cm^2, was the toxin detected in the second 2 mm cheese layer (7 ng/cm^2). Five of the six cheeses contained sterigmatocystin in the first 2 mm layer of the cheese itself (0·4–60 ng/cm^2). These observations led to use of analysis of the cheese scrapings only as a rapid screening procedure for detection of sterigmatocystin in hard cheese such as Gouda or Edam (Van Egmond *et al.*, 1982*c*). Distribution studies on sterigmatocystin in cheese are being continued in order to provide sufficient data for design of sampling procedures for future regulatory purposes (Van Egmond & Paulsch, 1986). In another migration study, no sterigmatocystin was detected in the centre part of three naturally-contaminated hard cheeses that contained 45–330 ng/g in the outer 2 cm layer (Lafont *et al.*, 1979*b*).

4.2.4 Aflatoxins

Studies on migration of aflatoxins in cheese have been carried out using Cheddar, Tilsit, Provolone, processed, Brick, and Feta cheeses.

Lie & Marth (1967) inoculated Cheddar cheeses on a freshly cut surface with spore suspensions of toxigenic strains of *Aspergillus flavus* and *A. parasiticus* and incubated them at room temperature. After 52 days, the upper 0·64 cm layer of cheese (after careful removal of mould mycelium) contained substantial levels of aflatoxins B$_1$ and G$_1$ (up to 15 600 ng/g of each), the second 0·64 cm layer contained up to 312 ng/g of either aflatoxin, while none was detected more than 1·3 cm from the cheese surface. The concentration of aflatoxins in the mycelium was higher than in the cheese.

Similar results to the above, but with lower aflatoxin levels in general, were obtained by Frank (1968), who incubated *A. flavus* on Tilsit cheese at 30°C for 6 days. In the first 0·5 cm slice, 200 ng/g of aflatoxin B$_1$ and 600 ng/g of aflatoxin G$_1$ were found, the second 0·5 cm slice contained 20 ng/g of each aflatoxin, but in the layers below no aflatoxins were detected. The same pattern was observed by Kiermeier & Behringer (1977*b*) in Tilsit cheese kept at 14°C for up to 52 days. However, in Provolone cheese incubated at 20°C, for up to 40 days, traces of aflatoxins B$_1$ and G$_1$ were still found in the layer 1–2 cm from the surface, although the levels of aflatoxins B$_1$ and G$_1$ were no greater than 16 ng/g in the 0·2–0·5 cm layer and 41 ng/g in the surface layer (Kiermeier & Behringer, 1977*b*).

Diffusion of aflatoxins has also been studied in processed cheese (Park & Bullerman, 1983). Samples were inoculated with *A. parasiticus* (poor production of aflatoxins) or *A. flavus* and aflatoxins B_1 and G_1 (totalling 96 ng/g) were detected in the second 1 cm layer after 21 days at 25°C, compared to a total of 1440 ng/g in the first 1 cm layer. Aflatoxins did not diffuse into the second layer at 15°C (21 days). In one sample of processed cheese inoculated along the top side of the punctured package with *A. parasiticus* and containing a mean concentration of 1600 ng/g of aflatoxins, diffusion of toxin through all eight slices of the cheese in the package was observed (Park & Bullerman, 1983).

Brick cheese allowed penetration of aflatoxins further than found in the above experiments with Cheddar, Tilsit, Provolone, and processed cheeses (Shih & Marth, 1972*a*). Aflatoxins B_1, B_2, G_1 and G_2 were produced by *A. parasiticus* at 24°C over 14 weeks and the fourth 1 cm layer contained 19 ng/g of aflatoxin B_1 (compared to 78 ng/g in the top 1 cm layer) and 79 ng/g aflatoxin G_1 (compared to 98 ng/g in the top 1 cm layer). Penetration of aflatoxins B_1 and G_1 into the fourth layer was also seen at 13°C after incubation of the cheese for 1 week; concentrations were much lower than at 24°C. It should be noted, however, that penetration of aflatoxins beyond 1 cm was variable and did not always occur at either temperature in samples analysed at different times or with *A. flavus* inoculated cheese kept at 24°C.

Aflatoxins B_1 and G_1 diffused into Feta cheese inoculated with *A. flavus* and kept at 13°C for 20 days or at 26°C for 7 days (Karaioannoglou, 1984). Maximum concentrations in the third 0·8 cm layer as a percentage of those in the top 0·8 cm were 0·9–1·5%.

Kiermeier & Behringer (1972) divided a block of Tilsit cheese which did not appear to be infected with mould but was naturally contaminated with aflatoxins B_1, B_2, G_1 and G_2 (mean B_1 7 ng/g) into eight 2 cm-slices; for each 9 cm × 13 cm slice they analysed a middle portion and six outer pieces. The concentrations in the pieces were very different, ranging from 0–50 ng/g for aflatoxin B_1 and 0–88 ng/g for total aflatoxins. Surprisingly, the slice that had the highest concentration of aflatoxin B_1 in the outer pieces had none in the middle while another slice had 33 ng/g of aflatoxin B_1 in the middle piece and only up to 20 ng/g in the outer layers. This variation leads to sampling problems for such cheeses.

4.3 Practical Conclusions

It is apparent that too little consistent data are available to answer the question of how much cheese one should trim off a block of cheese that has gone mouldy in order to avoid ingestion of mycotoxins, particularly if it was stored in a refrigerator and contaminated with *Penicillium*. The lowest incubation temperature used in any study on mycotoxin migration in cheese was 13°C and that was for aflatoxins, which cannot be studied by fungal incubation techniques at ≤7°C as they are not formed (Oldham *et al.*, 1971). The aflatoxin studies do illustrate the variation in depth of penetration with type of cheese. Experiments that result in extensive mould growth in order to form mycotoxins do not perhaps reflect the frequent reality of small colonies of mould occurring on cheese. In the latter situation, removal of 1·3 cm or more of cheese around and beneath the mould has been recommended (Bullerman, 1981) but obviously more research with reference to practical situations is needed. Olsen (1979) advised trimming of 0·64 to 1·3 cm off mouldy refrigerator-stored cheese as long as the consumer had had no problems with milk spoilage in that refrigerator. However, results obtained from analysis of certain cheeses naturally contaminated with aflatoxins (Kiermeier & Behringer, 1972) or ochratoxin A (Chapman *et al.*, 1983), where mycotoxins were found in the centre core, make one wonder if there is an answer to the question of how much cheese should be trimmed. The Health Protection Branch of Health and Welfare Canada (Anon., 1981) advises that if a hard cheese has developed a patch of mould growth that is confined to one area, the cheese can be salvaged by cutting away the mould to a depth of 2·5 cm. Chapman *et al.* (1983) recommended the removal of a layer 'considerably in excess' of 2 cm. These are both conservative recommendations.

4.4 Stability of Mycotoxins in Cheese

A knowledge of the stability of a mycotoxin in a food is of value in predicting chances of its occurrence. However, this can be misleading: patulin and penicillic acid are not particularly stable when added to cheese, yet have been detected naturally in various cheeses at levels up to 500 ng/g (see Section 5.2). Thus in Swiss cheese stored at 5°C the concentration of patulin decreased very rapidly during the first 12 h after addition, when 21% remained, and by 168 h it was down to 8% of the originally added concentration of 50 μg/g (Lieu & Bullerman, 1977). Penicillic acid was similarly unstable in Swiss cheese and 6% remained after 168 h. Patulin (50 μg/g) added to Cheddar cheese and stored at 5°C and 25°C for up to 48 h rapidly disappeared, and after 3 h only 18–31% of added patulin was recovered (Stott & Bullerman, 1976). The reaction of patulin with Cheddar cheese was not affected by storage temperature or by the presence of microorganisms (nonsterile cheese vs sterile cheese) and therefore appeared to be chemical in nature. The instability of patulin and penicillic acid in cheese has been attributed to reaction with amino acids and sulphydryl compounds (Bullerman, 1981).

Ochratoxin A appears to be fairly stable in Cheddar cheese (Bullerman, 1981) with 49% being recovered after 48 h storage at 25°C. There are no data on stability of citrinin in cheese.

In contrast to the three *Penicillium* mycotoxins discussed above, aflatoxins B_1 and G_1 were found to be very stable during storage in Swiss cheese for 168 h at 5°C (Lieu & Bullerman, 1977). Aflatoxin B_1 was also not destroyed on melting cheese at 88–135°C to make processed cheese (Kiermeier & Rumpf, 1975). The stability of sterigmatocystin in the rind layer of naturally-contaminated Gouda cheeses has been studied at −18°C, +4°C and +16°C, at which temperatures the concentration of toxin did not significantly decrease over a three month storage period (Van Egmond *et al.*, 1982*b*). These results confirm an earlier study on Gouda cheese inoculated with *Aspergillus versicolor* and stored for three months at −18°C (Northolt *et al.*, 1980).

Experiments have been carried out to try and determine if washing cheese will remove mycotoxins (Karaioannoglou, 1984). Feta cheese was contaminated with aflatoxins by diffusion from brine containing 176–399 ng/g of total aflatoxins B_1 and G_1 and the bottom 8 mm layer was washed with a six-fold volume of water. In three experiments, appreciable concentrations of aflatoxins remained in this layer of cheese — 4·5–91% for aflatoxin B_1 and 9·4–89% for aflatoxin G_1 of concentrations (23–155 ng/g of aflatoxin B_1 and 45–129 ng/g of aflatoxin G_1) found in the top 8 mm layer of the cheese, which was not washed. Frequent washing of Tilsit cheese inoculated with *A. flavus* and *A. parasiticus* appeared to cause removal of some aflatoxin from the surface layer (Kiermeier & Behringer, 1977*b*).

5 NATURAL OCCURRENCE OF MYCOTOXINS IN MOULDY CHEESE

5.1 General

Considering the large effort that has gone into identifying fungi that can contaminate cheese, surveys of cheese for mycotoxins themselves have been relatively few. Tables 7 and 8 show positive findings.

TABLE 7
Natural occurrence of *Penicillium* mycotoxins in mouldy cheese[a]

Location	Samples	Mycotoxin	Incidence	Level(s)/(ng/g)	Reference
USA	Swiss cheese	Penicillic acid	4/33	Up to 500	Bullerman (1976a)
France	Blue cheese	Penicillic acid	1/110	820	
France	Hard cheese	Penicillic acid	3/48	Up to 340 (outer layer)	
France	Semi-hard cheese (mouldy crust)	Penicillic acid	5/39	Up to 710	
France	Goat cheese	Penicillic acid	2/18	45 210	
France	Hard cheese	Patulin	1/48	90 (outer layer)	Lafont et al. (1976b)
France	Semi-hard cheese (mouldy crust)	Patulin	4/39	45–355	
France	Goat cheese	Patulin	1/18	30	
France	Hard cheese	Mycophenolic acid	4/48	0·01–1000 (outer layer)	
France	Semi-hard cheese (mouldy crust)	Mycophenolic acid	7/39	0·01–5000	

United Kingdom	Various cheeses (wholesale)	Ochratoxin A Citrinin	15/25 2/25	Up to 260 Up to 50	Chapman et al. (1983)
United Kingdom	Various cheeses (retail, domestic)	Ochratoxin A Citrinin	3/19 15/19	Up to 7 Up to 50	Jarvis (1983)
United Kingdom	Cheese	Citrinin	3 samples	?[b]	Jarvis et al. (1985)
USA	Cheese trimmings	Ochratoxin A	1 sample	?	Bullerman (1976b)
Yugoslavia	Edam cheese	Ochratoxin A	2/25	820, 1100	Škrinjar & Žakula (1985)
Poland	Cheesecake (mouldy)	Ochratoxin A	1 sample	1075	Piskorska-Pliszczyńska & Borkowska-Opacka (1984)
USA	Cream cheese	Penitrem A	1 sample	?	Richard & Arp (1979)
South Africa	Gouda/Cheddar (stored 1 year/5°C)	Cyclopiazonic acid	?	35 000–70 000	Lück & Wehner (1979)

[a] Excluding toxins commonly formed by *Penicillia* in mould-ripened cheese.
[b] ?, not stated.

TABLE 8

Natural occurrence of *Aspergillus* mycotoxins in mouldy cheese

Country	Samples	Incidence	Level(s) (ng/g)	Reference
		Sterigmatocystin		
The Netherlands	Gouda, Edam cheeses (surface layer)	9/39	5–600	Northolt *et al.* (1980)
The Netherlands	Gouda cheese (surface layers)	6 samples	?[a]	Van Egmond *et al.* (1982b)
The Netherlands	Gouda, Edam cheeses	11/67	Up to 9 000 (scrapings)	Northolt & Van Egmond (1982a)
Czechoslovakia	Various hard cheeses	3/66	7·5–17·5	Bartoš & Matyáš (1982)
France	Hard cheese	3/48	Up to 330 (outer layer)	Lafont *et al.* (1979b)
		Aflatoxin B$_1$ and/or G$_1$		
USA	Cheese trimmings	1 sample	?	Bullerman (1976b)
Federal Republic of Germany	Various cheeses	16/222	Traces (<10)	Kiermeier & Böhm (1971)

Federal Republic of Germany	Tilsit cheese	1/1	7 B_1 (9 total)	Kiermeier & Behringer (1972)
Federal Republic of Germany	Processed cheese	2/115	?	Kiermeier & Rumpf (1975)
France	Various cheeses	3 (?)/90 (St Nectaire)	?	Jacquet et al. (1970)
France	Pepper cheese	1 sample	'High'	Jacquet & Teherani (1974)
Romania	Cheese/rind	6/34	?	Galea & Bara (1976)
India	Cheese	6/26	8–15 B_1	Paul et al. (1976)
Tunisia	Cheese	1/248	1 B_1	Boutrif et al. (1977)
Egypt	Local Roquefort cheese	2/62	?	Naguib et al. (1983)
Egypt	Cheese	79/133	10–50	Mahrous et al. (1978)
		β-Nitropropionic acid		
USA	Cheeses	5/18	Traces	Iwasaki & Kosikowski (1973)

[a] ?, not stated or unclear.

5.2 *Penicillium* Toxins

In view of the predominance of penicillia on mould-contaminated cheeses, it is not surprising that known *Penicillium* mycotoxins have been detected in such cheeses. Table 7 shows positive findings only and in some cases concern is warranted. In particular, Lafont *et al.* (1979*b*) observed moderate incidence of penicillic acid, patulin, and mycophenolic acid in semi-hard French cheeses. High incidences of ochratoxin A and citrinin (but no patulin or penicillic acid) were found in domestic hard, soft and processed cheeses and various imported cheeses in the United Kingdom (Chapman *et al.*, 1983; Jarvis, 1983).

The occurrence of penitrem A in refrigerated mouldy cream cheese that was involved in the intoxication of two dogs was described by Richard & Arp (1979). *Penicillium crustosum* was isolated from the cheese. Although the concentration of penitrem A in the cream cheese was not determined, about 0·15 g of the cheese caused an acute neurologic syndrome and death on oral administration to mice while 30 g fed to a 10 kg Beagle dog produced a mild neurologic reaction (Arp & Richard, 1979).

The presence of unidentified mycotoxins in samples of mainly *Penicillium*-contaminated Edam cheese collected from markets in Yugoslavia was demonstrated by Škrinjar & Žakula (1985). Forty percent of cheese extracts were lethal to chicken embryos.

There have been several surveys of cheese for *Penicillium* mycotoxins that gave negative results. Ochratoxin A and citrinin were not found in a West German survey of 49 cheeses from various countries (Nowotny *et al.*, 1983*a*). Twenty of the cheeses were allowed to develop mould in the refrigerator but none of 22 mycotoxins, including luteoskyrin, mycophenolic acid, cyclopiazonic acid, penicillic acid and patulin in addition to ochratoxin A and citrinin, were detected (Nowotny *et al.*, 1983*b*). In a South African survey of 42 samples of normal, mouldy, and mould-ripened cheese, no ochratoxins, patulin, or cyclopiazonic acid were detected (Lück *et al.*, 1976*b*). Bullerman (1976*a*) analyzed 33 samples (11 brands) of Swiss cheese for sale in Nebraska, USA for ochratoxin A, citrinin, penicillic acid, patulin, and luteoskyrin, and other mycotoxins, with negative results except for penicillic acid (Table 7). Mouldy cheese trimmings from commercial outlets in the USA were negative for patulin and penicillic acid, but one sample contained ochratoxin A (Bullerman, 1976*b*). No ochratoxin A was detected in 24 samples of Canadian processed cheese (Williams, 1985) or in nine samples of mouldy East German cheese (Fritz & Engst, 1981; Fritz, 1983).

5.3 *Aspergillus* Toxins

The most concern regarding mycotoxins in cheese has been with sterigmatocystin (Table 8) (Northolt *et al.*, 1980; Van Egmond *et al.*, 1982*b*). In the upper 0·4–1 cm layers of warehouse-stored Edam and Gouda cheeses that were moulded with *Aspergillus versicolor* and varied in age from under two to more than seven months, sterigmatocystin was found in cheeses of all age categories. Retail samples of Edam Cake and Moravian Block cheeses contained low levels of sterigmatocystin in a Czechoslovakian survey (Bartoš & Matyáš, 1982). Sterigmatocystin was not detected in surveys for mycotoxins in cheese carried out in the United Kingdom (Steering Group on Food Surveillance, 1987), West Germany (Nowotny *et al.*, 1983*a,b*), South Africa (Lück *et al.*, 1976*b*), and the USA (Bullerman, 1976*a*; Stoloff, 1982). Just to mention the possibility of indirect contamination, although sterigmatocystin rarely occurs in animal feeds, it has been shown experimentally that carryover to cow's milk can occur and that the toxin transfers into cheese curd (Kiermeier & Kraus, 1980).

There is little documented evidence that any appreciable concentrations of aflatoxins B_1 or G_1 form naturally in cheese (Table 8) and in fact there have been many surveys reporting negative results, involving analysis of more than 1500 samples of various types of cheese in 15 countries (Frank & Eyrich, 1968; Boutibonnes & Jacquet, 1969; Shih & Marth, 1969; Jacquet *et al.*, 1970; Kiermeier & Groll, 1970*b*; Rothenbühler & Bachmann, 1970; Duitschaever & Irvine, 1971; Buliński, 1972; Hanssen & Jung, 1973; Bullerman, 1976*a*; Lück *et al.*, 1976*b*; Carini & Cerutti, 1977; Laub & Woller, 1977; Polzhofer, 1977*b*; Corbion & Frémy, 1978; Dulley & Houlihan, 1979; Lafont *et al.*, 1979*b*; Saito *et al.*, 1979; Fritz & Engst, 1981; Mashaly *et al.*, 1981; Goto *et al.*, 1982; Fonseca *et al.*, 1983; Fritz, 1983; Nowotny *et al.*, 1983*b*; Piskorska-Pliszczyńska & Borkowska-Opacka, 1984; Škrinjar & Žakula, 1985; Zerfiridis, 1985; Koch & Ross, 1986).

Traces of β-nitropropionic acid have been detected in cheese (Iwasaki & Kosikowski, 1973); another study was negative (detection limit about 3 μg/g) (Gilbert *et al.*, 1977).

5.4 Conclusions

Natural occurrence of the *Penicillium* mycotoxins ochratoxin A, citrinin, patulin, penicillic acid, penitrem A, and α-cyclopiazonic acid and the *Aspergillus* mycotoxins sterigmatocystin, aflatoxins B_1 and G_1, and β-nitropropionic acid has been demonstrated in mouldy cheese (other than intentionally mould-ripened cheese). These findings include moderate to high incidences of penicillic acid, patulin and mycophenolic acid in semi-hard French cheeses, ochratoxin A and citrinin in United Kingdom cheeses, sterigmatocystin in Dutch cheeses, and aflatoxins in some Egyptian cheeses. Nevertheless, there have been several surveys of cheese for *Penicillium* mycotoxins and sterigmatocystin and numerous surveys for aflatoxins that have yielded negative results. Further information would be desirable on the natural occurrence of penitrem A in *Penicillium*-contaminated cheese.

6 OTHER DAIRY PRODUCTS

6.1 Fungi Isolated from Dairy Products Other than Cheese

6.1.1 Milk

Fungi of the following genera have been isolated from milk (usually denoted as raw milk) that was examined in the Federal Republic of Germany, Yugoslavia, the USA, Egypt, and Japan: *Mucor, Fusarium, Geotrichum, Penicillium, Alternaria, Cladosporium, Aspergillus* and *Rhizopus*, although the most frequently isolated fungi varied with the study (Cooke & Brazis, 1968; Texdorf, 1972; Nakae & Yoneya, 1976; Šutić *et al.*, 1979; Megalla *et al.*, 1981; Škrinjar *et al.*, 1983; Pitt & Hocking, 1985). *Aspergillus flavus* was one of the most commonly found species in market milk in Assiut, Egypt and 2 out of 28 isolates were aflatoxigenic (Megalla *et al.*, 1981); *A. flavus* was also detected in raw milk in Yugoslavia (Škrinjar *et al.*, 1983) and Japan (Nakae & Yoneya, 1976). Six of 39 strains of fungi isolated from the Japanese milk were toxic to the chick embryo — these were *Aspergillus candidus, A. niger, A. ochraceus, A. terreus, Alternaria* sp., and *Mucor* sp.

6.1.2 Dried milk products

A number of studies have indicated the importance of milk powder as a potential source of toxigenic fungi. The major fungal genera found have been *Penicillium, Cladosporium, Alternaria, Mucor,* and *Aspergillus* (Boutibonnes & Jacquet, 1969; Moreau, 1976*a*; Jesenská & Poláková, 1978; Aboul-Khier *et al.*, 1985). *Aspergillus* spp. known to be mycotoxigenic predominated in studies carried out in Egypt and France (Moreau, 1976*a*; Aboul-Khier *et al.*, 1985) and *A. flavus* was frequently detected. Isolations of *A. flavus* from powdered milk have also been reported by Boutibonnes & Jacquet (1969), Jesenká & Poláková (1978), and Šutić *et al.* (1979).

Potentially mycotoxigenic *Penicillium* species — *P. frequentans, P. cyclopium, P. decumbens,* and *P. expansum* — predominated in some Spanish milk-cereal products for infants (de Simón & Fernandez, 1983); *Cladosporium cladosporioides, P. cyclopium, Aspergillus glaucus,* and *A. versicolor* were the main fungi found in spray milk (Le Bars, 1985); and *A. flavus, A. niger, A. fumigatus, Penicillium* spp., *Cladosporium* spp., and *Rhizopus* spp. were among the most frequently isolated fungi from Egyptian dried ice cream mixes (Aboul-Khier *et al.*, 1985).

6.1.3 Cream and butter

Spoilage of fresh cream by *Geotrichum candidum, Aspergillus glaucus* (in particular *A. amstelodami*), and *Penicillium stecki* has been reported (Marth, 1978; Moreau, 1983). *Phoma* species have also been isolated from cream (Pitt & Hocking, 1985). A wide variety of moulds from genera known to include toxigenic species have been found on butter, namely *Cladosporium* (*Hormodendrum*), *Alternaria, Aspergillus, Mucor, Fusarium, Phoma, Scopulariopsis, Paecilomyces, Rhizopus, Penicillium, Verticillium,* and *Geotrichum* (Sasaki, 1950; Foster *et al.*, 1957; Inagaki, 1962; Muys *et al.*, 1966; Thomas & Druce, 1971; Asad & Husain, 1972; Kozareva *et al.*, 1972; Ölsanský *et al.*, 1979; Moreau, 1983; Abdel-Rahman & El-Bassiony, 1984; Pitt & Hocking, 1985). *Cladosporium butyri* is a well known cause of rancidity in butter and cream (Muys *et al.*, 1966) and *Aspergillus flavus* has been recorded as fairly common on butter in certain regions of Japan (Sasaki, 1950).

6.1.4 Fermented milk products

Fungi isolated from yoghurt include *Penicillium, Monilia, Cladosporium, Geotrichum,* and *Mucor* (Dubois, 1981; Spillmann & Geiges, 1983; García & Fernandez, 1984). The *Penicillium* isolates included several known toxigenic species. *Penicillium* was also the main fungal genus isolated from fermented milk products in Yugoslavia (Šutić *et al.*, 1979).

Fungi may contaminate yoghurt from added fruit and nuts and in a

survey of 315 such yoghurts from various United Kingdom manu-
facturers, viable moulds were isolated from 28% of the samples, although
only 3% supported visible mould growth on prolonged storage (Jarvis,
1972). In another microbiological study of yoghurt in the United
Kingdom, mould counts were low or nil in both natural and fruit
yoghurts (Davis & McLachlan, 1974).

There appear to be no specific reports of *Aspergillus flavus* in either
yoghurt or buttermilk.

6.2 Production of Mycotoxins in Dairy Products Other than Cheese

6.2.1 Aflatoxins

Sterilized, homogenized milk (2·6% fat) is a favourable substrate for
production of aflatoxins by *Aspergillus flavus* (Boutibonnes & Jacquet,
1967). Total aflatoxin yields of up to 700 μg/g were obtained after nine
days at an incubation temperature of 22°C. Production of aflatoxins B_1
and G_1 by *A. flavus* on milk was also shown by Zappavigna & Cerutti
(1973). After just 24 h incubation of homogenized milk with *A.
parasiticus*, 200 ng total aflatoxin/ml was detected by Bennett *et al.*
(1980). Aflatoxins were stable for at least four days in sterilized milk
acidified with lactic acid; milk seems to prevent the acid transformation
of aflatoxins to more polar products (which were not identified but are
presumably aflatoxins B_{2a} and G_{2a}) (Jacquet & Boutibonnes, 1970).

Cream is another excellent substrate for aflatoxigenesis and a
concentration of 520 μg/ml of aflatoxins B_1 + B_2 from *A. flavus* has been
reported (Jacquet *et al.*, 1970; Jacquet & Boutibonnes, 1970). Good
yields of aflatoxin B_1 (up to 78 μg/ml) were also obtained when *A. flavus*
was grown on milk enriched with 20% cream (Jacquet & Tantaoui-
Elaraki, 1976).

Production of aflatoxin B_1 on milk powder was first demonstrated by
Frank (1966) using *A. flavus*. Kiermeier (1971*b*) studied the influence of
water content and relative humidity on aflatoxin production at 20°C by
A. parasiticus and *A. flavus* in milk powder; up to 192 μg/g of aflatoxins B_1
and G_1 were found at 60% water content but none was formed below 50%
water content. Of particular interest was the formation of up to 2·1 μg/g
of aflatoxins B_1 and G_1 (mainly B_1) by *Penicillium puberulum* in moist
milk powder. Later, detailed studies with moist milk powder inoculated
with *A. parasiticus* and incubated at temperatures of 1–20°C for 14 days

showed that 87·2 µg/g of aflatoxin G_1 and smaller concentrations of aflatoxins G_2, B_1, B_2 and M_1 were formed at 20°C, while even at 1°C, 0·58 µg/g of aflatoxin G_1 was detected (Kiermeier & Behringer, 1977a). Control milk powder analysed without incubation also contained low levels of aflatoxins (up to 0·052 µg G_1 per g) indicating contamination from the fungal spores. Preincubation at 20°C for three days greatly increased the concentration of aflatoxins B_1 and G_1 formed at 10°C by *A. parasiticus*. In other model experiments with moist milk powder inoculated with *A. flavus*, raising the initial pH in the range 4·5–7·5 correspondingly lowered the amounts of aflatoxins B_1 and G_1 produced (Kiermeier & Behringer, 1973).

Sweetened condensed milk does not support growth of *A. flavus* but unsweetened condensed milk does, with formation of aflatoxins (Frank, 1966; Jacquet & Boutibonnes, 1970; Jacquet & Tantaoui-Elaraki, 1976). Production of aflatoxins B_1 and G_1 by *A. parasiticus* and *A. flavus* on khoa, an Indian concentrated milk product with doughlike consistency, reached 81–92 µg/g (in the top 1 cm layer) after 14 days at 28°C (Lembhe *et al.*, 1981); penetration of the aflatoxins to the fourth 1-cm layer from the top was detected.

In milk-based infant formulae inoculated with *A. parasiticus* and incubated at 30°C for 24 h, aflatoxins B_1 and B_2 were produced at a level of 273 ng/ml (Bennett *et al.*, 1980).

While Frank (1966) found no growth of an *A. flavus* strain on butter after two weeks, Rehm & Schmidt (1969) observed very slow development of three strains over four months and production of low concentrations of aflatoxin B_1 (up to 0·5 µg/g) by two of these. Similarly, Jacquet & Tantaoui-Elaraki (1976) found very little or no growth of *A. flavus* on butter and formation of trace quantities only of aflatoxins.

Production of total aflatoxins on yoghurt by *A. flavus* at 25°C reached 138 µg/g after 21 days incubation (Park & Bullerman, 1983). However, aflatoxin B_1 was detoxified during fermentation of milk to yoghurt, forming the water addition product aflatoxin B_{2a} (Megalla & Hafez, 1982). Aflatoxin B_1 in milk fermented with *Streptococcus lactis* was transformed to aflatoxicol as well as aflatoxin B_{2a} (Megalla & Mohran, 1984).

6.2.2 Other mycotoxins

Curdled milk was found to be a suitable medium for production of cyclopiazonic acid by *Penicillium verrucosum* var. *cyclopium* (up to 80 µg/ml) (Malik *et al.*, 1986).

6.3 Natural Occurrence of Mycotoxins in Dairy Products Other than Cheese

The occurrence of aflatoxins in other dairy products is rare (Table 9). Negative surveys of milk and milk products other than cheese for aflatoxins B_1, B_2, G_1 and G_2 have involved analysis of over 1500 samples in 12 countries (Frank & Eyrich, 1968; Boutibonnes & Jacquet, 1969; Jacquet *et al.*, 1970; Yndestad & Underdal, 1975; Lück *et al.*, 1976a; Paul *et al.*, 1976; Boutrif *et al.*, 1977; Polzhofer, 1977b; Dulley & Houlihan, 1979; Saito *et al.*, 1979; Fritz & Engst, 1981; Mashaly *et al.*, 1981; Goto *et al.*, 1982; Fritz, 1983; Škrinjar *et al.*, 1983; Villarejo *et al.*, 1984). It should be pointed out too that aflatoxin B_1 has been detected in animal milk after experimental dosing of goats, cows and a ewe with aflatoxin B_1 or mixed aflatoxins, although concentrations were very much lower than for aflatoxin M_1 (Nabney *et al.*, 1967; Masri *et al.*, 1969; Jacobson *et al.*, 1971; Ferrando *et al.*, 1977; Veselý *et al.*, 1978).

Recently, ochratoxin A was detected at a level of 6·8 ng/g in milk powder associated with deaths of Afghan puppies. The deaths were probably caused by a virus (Gareis *et al.*, 1987). A survey of 22 unmouldy milk products other than cheese for ochratoxin A in the German Democratic Republic was negative (Fritz & Engst, 1981; Fritz, 1983).

6.4 Conclusions

Mucor, Fusarium, Geotrichum, Penicillium, Alternaria, Cladosporium, Aspergillus and *Rhizopus* are the major genera of potentially toxigenic fungi isolated from milk and some of these have been found in milk products such as powdered milk and yoghurt; in addition to these genera, *Scopulariopsis* and *Verticillium* have been isolated from butter. Milk and cream are excellent substrates for experimental production of aflatoxins by *Aspergillus flavus* while moist milk powder, unsweetened condensed milk, and yoghurt are also good substrates, but butter is not. Very little is known about production of other mycotoxins on dairy products other than cheese. The natural occurrence of aflatoxins in other dairy products due to fungal contamination is rare; although a few notable findings in European milk powder and milk have been reported, there have been numerous negative surveys. Ochratoxin A has only been detected in one sample of milk powder.

TABLE 9

Natural occurrence of aflatoxins B_1, B_2, G_1 and G_2 in milk and milk products other than cheese

Country	Samples	Aflatoxins			Reference
		Incidence	*Level(s) (ng/g)*		
German Democratic Republic	Milk powder (dairy) products	1/18 (22) (infant formula)	6·4 B_1		Fritz *et al.* (1977); Fritz & Engst (1981); Fritz (1983)
Czechoslovakia	Leftover milk powder (associated with aflatoxin B_1 in the liver)	5/5	42–550 B_1		Jesenská & Poláková (1978)
	Milk powder (as above, packaged)	4 samples	320–5400 B_1 in surface layers		
Spain	Natural milk	2/1150	0·28–0·36 B_1		Villarejo *et al.* (1984)
Yugoslavia	Milk	5/105	Up to 2 500 (mainly B_1)		Šutić *et al.* (1979)
France	Milk powder	3·6%	?[a]		Jacquet & Lafont (1979)
Iran	Milk	2/95	?		Suzangar *et al.* (1976)
India	Indigenous milk products	2/23	10, 20 B_1		Paul *et al.* (1976)
Japan	Milk, milk powder, evaporated milk	1/320	682 B_1, +24 B_2		Toya (1985)

[a] ?, not stated.

7 ANALYTICAL METHODS FOR MYCOTOXINS WITH POTENTIAL TO CONTAMINATE DAIRY PRODUCTS DIRECTLY

7.1 General

Although in a few investigations on the natural occurrence of mycotoxins in cheese, use was made of methods that were available for other commodities, several methods have been developed especially for cheese and other dairy products. These are particularly applicable for sterigmatocystin and aflatoxin analysis. Most methods used for determination of mycotoxins in cheese are thin-layer chromatographic (TLC) methods.

7.2 Methods for *Penicillium* Mycotoxins (Multimycotoxin Methods)

A number of multimycotoxin methods that include various *Penicillium* toxins have been developed, most of them applicable to grains and feeds (Steyn, 1981); some have been applied to cheese. Extraction and cleanup procedures vary. Where TLC has been used for detection, two-dimensional development is often favoured, or different TLC systems are used unidimensionally, and the mycotoxins are visualized in various ways — usually with spray reagents and/or fluorescence under ultraviolet light. Reverse phase high performance liquid chromatography (HPLC) has been used in some cases.

A method was developed by Siriwardana & Lafont (1979) for determination of mycophenolic acid, patulin and penicillic acid, sterigmatocystin and aflatoxins, specifically in cheese. The cheese was extracted with 5% sodium chloride–methanol–acetone (1:1:1, v/v/v) containing acetic acid, caseins were precipitated at −25°C and, after a defatting step using hexane, the mycotoxins were transferred to chloroform and ethyl acetate and the extract was further purified by column chromatography. Two-dimensional TLC was used for the first three mycotoxins yielding detection limits of 20, 20 and 30 ng/g, respectively.

Ochratoxin A and citrinin (also sterigmatocystin) were determined in cheese by Nowotny *et al.* (1983*a*). An acidic methanol extract was treated with Carrez solution, defatted with hexane, and extracted with dichloromethane. Limits of detection for ochratoxin A and citrinin were respectively 20 and 10 ng/g by HPLC and 40 and 10 ng/g by TLC and recoveries of citrinin measured by HPLC were much better than by TLC. The same research group developed a similar method to detect 22 mycotoxins in mouldy cheese and other foods (Nowotny *et al.*, 1983*b*); *Penicillium* toxins included were ochratoxin A, citrinin, luteoskyrin, mycophenolic acid, cyclopiazonic acid, penicillic acid and patulin, with detection limits of 50 to 1000 ng/g.

Cheese samples were analysed by Chapman *et al.* (1983) for ochratoxin A, citrinin, patulin, penicillic acid and also sterigmatocystin by a method incorporating a hexane defatting step, chloroform extraction, a membrane dialysis cleanup (Roberts & Patterson, 1975) and TLC. Subsequently, ochratoxin A and citrinin were determined by modifying the HPLC method of Wilson *et al.* (1976), which involves extraction with chloroform–0·1M phosphoric acid (10:1, v/v) and cleanup on a basic Celite column. Detection limits were approximately 20 ng/g and 50 ng/g for ochratoxin A and citrinin, respectively (Chapman *et al.*, 1983; Jarvis, 1983).

The method used by Bullerman (1976*a*) to detect penicillic acid in cheese was a modification of a method for patulin in grains (Pohland & Allen, 1970) — 3% formic acid was added to the acetonitrile extraction solvent — and included ochratoxin A, citrinin, patulin, and luteoskyrin (in addition to aflatoxins, sterigmatocystin and zearalenone).

Methodology for α-cyclopiazonic acid, mycophenolic acid and various *Penicillium* alkaloids has been developed with regard to mould-ripened cheeses (see Chapter 6).

Few methods have been published concerning analysis of milk and milk products other than cheese for *Penicillium* mycotoxins. Gertz & Böschemeyer (1980) developed a TLC screening method for ochratoxin A, penicillic acid, patulin, sterigmatocystin and the aflatoxins in cheese, milk and other foodstuffs, which were extracted with ethyl acetate and cleaned up on a small silica gel column. Powdered milk and whey were substrates included in a method for determining ochratoxin A in feeds and feed ingredients (Jizdny, 1983).

7.3 Sterigmatocystin

Sterigmatocystin is usually detected on the TLC plate by spraying and heating with aluminium chloride solution. Bright yellow fluorescent spots are seen under longwave ultraviolet (UV) light (Van Egmond *et al.*, 1980; Francis *et al.*, 1985). This fluorescence could be enhanced 10-fold by additional spraying of the plate with a silicone–ether mixture to give detectability of 1–3 ng on the TLC plate (Francis *et al.*, 1985). Spraying with paraffin-*n*-hexane similarly enhanced zirconyl chloride produced fluorescence of sterigmatocystin by 20-fold (1 ng detectable) (Gertz & Böschemeyer, 1980). The identity of sterigmatocystin may be confirmed by a two-dimensional TLC test based on reaction with trifluoroacetic acid after the first development of the plate (Van Egmond *et al.*, 1980). The overall methods use methanol–4% KCl (9:1, v/v) or acetonitrile–4% KCl (85:15, v/v) for extraction of cheese followed by Florisil® and polyamide column cleanup steps (Van Egmond *et al.*, 1980, 1982*a*) or liquid–liquid partition and a cupric carbonate column (Francis *et al.*, 1985). Bartoš & Matyáš (1982) omitted the polyamide column step in the method of Van Egmond *et al.* (1980). Sterigmatocystin can be detected at levels as low as 2 ng/g cheese by the one-dimensional TLC procedure of Francis *et al.* (1985), whose method has undergone interlaboratory collaborative study (Francis *et al.*, 1987), or 5 ng/g cheese by the two-dimensional TLC method (Van Egmond *et al.*, 1980; Bartoš & Matyáš, 1982). Analysis of cheese scrapings by two-dimensional TLC was developed as a rapid screening method for sterigmatocystin in hard cheese, which, with no column cleanup, had a limit of detection of 100 ng/g; this was considered suitable for the purpose and required a sample size of only 50–100 mg (Van Egmond *et al.*, 1982*c*).

A multimycotoxin TLC method for determination of several myco-toxins in cheese had a detection limit of 20 ng/g for sterigmatocystin (Siriwardana & Lafont, 1979) and sterigmatocystin has been included in other multimycotoxin methods suitable for cheese (Bullerman, 1976*a*; Gertz & Böschemeyer, 1980; Chapman *et al.*, 1983; Nowotny *et al.*, 1983*b*) (Section 7.2). In one of these methods, milk was also an analytical substrate (Gertz & Böschemeyer, 1980). A reverse phase HPLC method for determination of sterigmatocystin, citrinin and ochratoxin A in cheese was described by Nowotny *et al.* (1983*a*). Sterigmatocystin was detected by UV absorption at 246 nm and the detection limit was 20 ng/g compared to 50 ng/g for TLC. However, recoveries were ≤60%.

7.4 Aflatoxins B₁, B₂, G₁ and G₂

Many methods have been used for determination of aflatoxins in cheese and other dairy products — perusal of the references in Tables 8 and 9 and in Sections 5.3 and 6.3 will give some of them.

Rothenbühler & Bachmann (1970) defatted cheese with petroleum ether then extracted it with acetone–water (7:3, v/v). Aqueous lead acetate was used for cleanup and TLC with several solvent systems for detection of the aflatoxins. Saito *et al.* (1979) also extracted aflatoxins from cheese and other dairy products with acetone–water (10:3, v/v). After several cleanup steps the aflatoxins, including M_1, were determined by TLC with fluorodensitometry.

Shih & Marth (1971) extracted aflatoxins from cheese and other foods with chloroform, methanol, and water in such proportions that a monophasic system was formed; chloroform and water were then added to give a biphasic mixture, which was filtered and then the chloroform layer was separated. Purification was by two solvent partitions and determination was carried out by TLC. Corbion & Frémy (1978) modified this method slightly for their survey of Camembert cheese.

Kiermeier & Weiss (1976) published a simple and rapid densitometric TLC method for aflatoxins B_1, B_2, G_1, G_2 and M_1 in milk, cheese, and other milk products. Chloroform was used for extraction and aqueous sulphosalicylic acid was added to break up any emulsion. Cleanup was by chromatography on a silica gel column. Emulsions formed during extraction of processed cheese were broken with urea solution (Kiermeier & Rumpf, 1975). The same laboratory had previously published other methods for analysis of aflatoxins in cheese (Kiermeier, 1970; Kiermeier & Groll, 1970*a*) and used the chicken embryo test for confirmation of identity (down to the 10 ng/g level), as many false positives were observed by TLC at trace levels (Kiermeier & Böhm, 1971).

A sensitive HPLC method for aflatoxins B_1, B_2, G_1, G_2 and M_1 and M_2 in cheese, milk and butter was developed by Goto *et al.* (1982). Toluene–ethyl acetate–formic acid–methanol (90:5:2·5:2·5, v/v/v/v) was used as mobile phase for normal phase HPLC because all the separated aflatoxins can be readily detected by fluorescence in this system.

At the time of writing, the most recent method developed for determination of aflatoxins in cheese is that of Kamimura *et al.* (1985). Extraction was with chloroform–water and extracts were purified on a Florisil® column before determination by fluorodensitometric high performance TLC. The method is applicable to aflatoxin B_1, B_2, G_1, G_2 and M_1 and to other foods and the minimum detectable aflatoxin concentrations were 0·1–0·2 ng/g.

For screening purposes, a minicolumn method for detecting aflatoxins B_1 and G_1 in cheese and other dairy products down to levels of 5 ng/g was developed by Metwally *et al.* (1983).

Analytical methods specifically for determination of aflatoxins in milk and milk products other than cheese have also been published (Jacobson *et al.*, 1971; Fritz *et al.*, 1977; Toya, 1985; Adensam *et al.*, 1986). Methods of Zimmerli (1977), Toya (1985) and Adensam *et al.* (1986) use HPLC and detection limits as low as 0·1 ng/g are attainable. Radio-immunoassay has also been applied to determination of aflatoxin B_1 (and M_1) in powdered milk, whey and buttermilk (Březina *et al.*, 1986).

7.5 β-Nitropropionic Acid

A gas chromatographic method for β-nitropropionic acid based on electron capture detection of its pentafluorobenzyl derivative has been applied to cheese by Gilbert *et al.* (1977). The limit of detection was of the order of 3000 ng/g. No β-nitropropionic acid was detected in various cheeses analyzed, in contrast to a previous study (Iwasaki & Kosikowski, 1973) where traces were found on five cheese surfaces using the less sensitive (and probably less specific) colourimetric procedure of Matsumoto *et al.* (1961).

7.6 Conclusions

There is no lack of methods for determination of aflatoxins, sterigmato-cystin and most of the *Penicillium* mycotoxins that might be expected to occur in mould-contaminated cheese. Nevertheless more sensitive methods for some of the *Penicillium* mycotoxins, in particular for penitrem A, would be desirable.

8 SUMMARY

The most common moulds that contaminate cheese are *Penicillium* species, which readily grow on cheese, even in the refrigerator. These can produce the mycotoxins ochratoxin A, citrinin, penicillic acid, patulin, mycophenolic acid, penitrem A and cyclopiazonic acid, all of which have been detected in cheese that has become mouldy. Especially significant are cheeses allowed to develop a mouldy crust during ripening. *Aspergillus* species have been found in appreciable incidence on cheeses in The Netherlands, where presence of sterigmatocystin in the cheese was associated with growth of *A. versicolor*. *A. flavus*, which produces aflatoxins, occurs extremely rarely on cheese and in any case does not generally grow and produce aflatoxins in cheese at refrigerator temperatures. Nevertheless, most of the research on cheese moulds and mycotoxins has concentrated on aflatoxigenic *Aspergillus* spp. and the aflatoxins. Penetration of aflatoxins into cheese from mouldy surface appears to vary with the type of cheese and can reach as far as 4 cm. Too few experiments on migration of *Penicillium* mycotoxins into cheese have been done. Patulin and penicillic acid are not particularly stable in cheese in contrast to the aflatoxins and sterigmatocystin.

Methods of analysis have been developed for determination of mycotoxins in cheese and (nearly all for aflatoxins only) in milk and other milk products. In most of the methods use is made of thin-layer chromatography for the final determination. These methods of course are not generally available to the consumer who must decide whether to discard or trim a piece of mouldy cheese.

REFERENCES

Abdel-Rahman, H. A. & El-Bassiony, T. (1984). Psychrotropic moulds in some food products. *Assiut Vet. Med. J.,* **13,** 135–43; *Fd Sci. Technol. Abstr.,* **18** (1986), no. 11 B 10.

Aboul-Khier, F., El-Bassiony, T., Hamid, A., A.-L. & Moustafa, M. K. (1985). Enumeration of molds and yeasts in dried milk and ice-cream products. *Assiut Vet. Med. J.,* **14,** 71–8; *Fd Sci. Technol. Abstr.,* **18** (1986), no. 7 P 9.

Adensam, L., Lebedová, M. & Turek, B. (1986). The determination of very low concentrations of aflatoxins. *Česk. Hyg.,* **31,** 282–7 (in Czech).

Andres, C. (1982). Mold/yeast inhibitor gains FDA approval for cheese. *Fd Process. USA,* **43,** 83.

Anon. (1981). *Mould — More Than Meets the Eye.* Health Protection Branch, Health and Welfare Canada, Ottawa.

Applebaum, R. S., Brackett, R. E., Wiseman, D. W. & Marth, E. H. (1982). Aflatoxin: toxicity to dairy cattle and occurrence in milk and milk products — a review. *J. Fd Protect.,* **45,** 752–77.

Arp, L. H. & Richard, J. L. (1979). Intoxication of dogs with the mycotoxin penitrem A. *J. Am. Vet. Med. Ass.,* **175,** 565–6.

Asad, F. & Husain, S. S. (1972). Fungi from butter. *Sci. Ind.,* **9,** 287–8.

Augusto, G., Luigi, M. P. & Adriano, C. (1968). Microflora of Robbiola cheese. *Latte,* **42,** 871–3 (in Italian); *Dairy Sci. Abstr.,* **32** (1970), no. 3453.

Bachmann, M., Lüthy, J. & Schlatter, C. (1979). Toxicity and mutagenicity of molds of the *Aspergillus glaucus* group. Identification of physcion and three related anthraquinones as main toxic constituents from *Aspergillus chevalieri. J. Agric. Fd Chem.,* **27,** 1342–7.

Bartoš, J. & Matyáš, Z. (1982). Examination of cheeses for sterigmatocystin. *Vet. Med. (Prague),* **27,** 747–52 (in Czech).

Bennett, J. W., Dunn, J. J. & Goldberg, E. J. (1980). Infant formulas as substrates for aflatoxin production. *Dev. Ind. Microbiol.,* **21,** 379–83.

Betina, V. (Ed.) (1984). *Developments in Food Science, Vol. 8. Mycotoxins. Production, Isolation, Separation and Purification.* Elsevier, Amsterdam.

Bilai, V. I. (1960). *Mycotoxins of Man and Agricultural Animals.* US Department of Commerce, Washington, DC.

Bodini, T., Guicciardi, A. & Craveri, R. (1969). Microbiological and chemical study of some soft cheeses, and the identification of the yeasts and fungi. *Latte,* **43**, 711–19 (in Italian); *Dairy Sci. Abstr.,* **32** (1970), no. 3452.

Boer, E. de & Stolk-Horsthuis, M. (1977). Sensitivity to natamycin (pimaricin) of fungi isolated in cheese warehouses. *J. Fd Protect.,* **40**, 533–6.

Boersma, H. Y. (1979). Mould growth on cheese, its consequences and control in practice. *Neth. Milk Dairy J.,* **33**, 208–10.

Boutibonnes, P. & Jacquet, J. (1967). Recherches sur la production de toxine par *Aspergillus flavus. Bull. Acad. Vét. Fr.,* **40**, 393–403.

Boutibonnes, P. & Jacquet, J. (1969). Sur la fréquence de l'aflatoxine et des *Aspergillus flavus* dans les aliments. *CR Soc, Biol.,* **163**, 1119–24.

Boutrif, E., Jemmali, M., Campbell, A. D. & Pohland, A. E. (1977). Aflatoxin in Tunisian foods and foodstuffs. *Ann. Nutr. Alim.,* **31**, 431–4.

Brandl, E. (1976). On toxic mould metabolites in milk and milk products. *Wien. Tierärztl. Monatschr.,* **63**, 166–71 (in German).

Brenet, M., Centeleghe, J. L., Millière, J. B., Ramet, J. P. & Weber, F. (1972). Étude d'un accident en fromagerie de type 'Camembert' causé par des mucorales. *Lait,* **52**, 141–8.

Březina, P., Švecova, T., Rausch, P., Lohniský, J. & Svobodova, M. (1986). The problem of aflatoxins determination. *Krmivarstvi Sluzby,* **22**, 175–7 (in Czech); *Chem. Abstr.,* **105**, 224721k.

Broquist, H. P. (1986). Slaframine and swainsonine, mycotoxins from *Rhizoctonia leguminicola. J. Toxicol. — Toxin Rev.,* **5**, 241–52.

Buliński, R. (1972). Examination for aflatoxins of moulds used in cheesemaking and of mould-ripened cheeses. *Roczn. Inst. Przem. Mlecz.,* **14**, 35–41 (in Polish); *Dairy Sci. Abstr.,* **34** (1972), no. 5133.

Bullerman, L. B. (1976*a*). Examination of Swiss cheese for incidence of mycotoxin producing molds. *J. Fd Sci.,* **41**, 26–8.

Bullerman, L. B. (1976*b*). Toxinogenic potential of molds isolated from moldy cheese trimmings. *J. Milk Fd Technol.,* **39**, 705.

Bullerman, L. B. (1977*a*). Incidence and control of mycotoxin producing molds in domestic and imported cheeses. *Ann. Nutr. Alim.,* **31**, 435–46.

Bullerman, L. B. (1977*b*). Mold control in the cheese plant. *Ital. Cheese J.,* **6**, 1–8.

Bullerman, L. B. (1980). Incidence of mycotoxic molds in domestic and imported cheeses. *J. Fd Safety,* **2**, 47–58.

Bullerman, L. B. (1981). Public health significance of molds and mycotoxins in fermented dairy products. *J. Dairy Sci.,* **64**, 2439–52.

Bullerman, L. B. & Olivigni, F. J. (1974). Mycotoxin producing-potential of molds isolated from Cheddar cheese. *J. Fd Sci.,* **39**, 1166–8.

Bullerman, L. B., Schroeder, L. L. & Park, K.-Y. (1984). Formation and control of mycotoxins in food. *J. Fd Protect.,* **47**, 637–46.

Burbianka, M. & Stec, E. (1972). Fungal contamination of the domestic food-stuffs. *Rocz. Panst. Zakl. Hig.,* **23**, 41–7 (in Polish).

Burzyńska, H. (1977). Possible production of aflatoxin B_1 in mayonnaise and cottage cheese. *Zesz. Probl. Postępów Nauk Roln.,* **189**, 83–7.

Carini, S. & Cerutti, G. (1977). Moulds and mycotoxins in Italian cheeses. *Ind. Alim.,* **16**, 106–8 (in Italian).

Chapman, W. B., Cooper, S. J., Norton, D. M.., Williams, A. R. & Jarvis, B. (1983). Mycotoxins in molded cheeses. In *Proceedings of the International Symposium on Mycotoxins,* Cairo, Egypt, 6–8 September 1981, ed. K. Naguib, M. M. Naguib, D. L. Park & A. E. Pohland. Food and Drug Administration, Rockville, Maryland, USA and National Research Centre, Cairo, Egypt, pp. 363–73.

Chattopadhyay, S. K., Nandi, B., Ghosh, P. & Thakur, S. (1987). A new mycotoxin from *Aspergillus candidus* link isolated from rough rice. *Mycopathologia,* **98**, 21–6.

Cole, R. J. & Cox, R. H. (1981). *Handbook of Toxic Fungal Metabolites.* Academic Press, New York.

Cooke, W. B. & Brazis, A. R. (1968). Occurrence of molds and yeasts in dairy products. *Mycopath. Mycol. Appl.,* **35**, 281–9.

Cooper, S. J., Wood, G. M., Chapman, W. B. & Williams, A. P. (1982). Mycotoxins occurring in mould damaged foods. In *Proc. V Int. IUPAC Symp. on Mycotoxins and Phycotoxins,* 1–3 September 1982, Vienna. Austrian Chemical Society, Vienna, 1982, pp. 64–7.

Corbion, B. & Frémy, J. M. (1978). Recherche des aflatoxines B_1 et M_1 dans les fromages de type 'Camembert'. *Lait,* **58**, 133–40.

Dale, G. (1972). Moulds and yeasts in the flora of Saint-Nectaire cheese. *Rev. Lait. Fr.,* 199, 201, 203 (in French); *Dairy Sci. Abstr.,* **34** (1972), no. 5132.

Dale, G. & Guillot, J. (1971). Contribution à l'étude de la microflore du fromage de Sainte-Nectaire et de la physiologie du *Geotrichum candidum* Link. *CR Séanc. Soc. Biol.,* **165**, 309–16.

Davis, J. G. & McLachlan, T. (1974). Yogurt in the United Kingdom: chemical and microbiological analysis. *Dairy Ind.,* **39**, 149–57, 177.

Delespaul, G., Gueguen, M. & Lenoir, J. (1973). La flore fongique superficielle des fromages de St-Nectaire et de tome de Savoie — son évolution au cours de l'affinage. *Rev. Lait. Fr., No. 313,* October, pp. 1–8.

Demirer, M. A. (1974). A study of moulds isolated from certain cheeses and their ability to produce aflatoxins. *Ankara Üniversitesi Veteriner Fakültesi Dergisi,* **21**, 180–98 (in Turkish); *Dairy Sci. Abstr.* **38** (1976), no. 4981.

Desfleurs, M. (1975). Contamination de Pont-l'Evêque par un *Penicillium* provenant des boîtes et cageots servant à leur emballage. *Lait,* **55**, 396–400.

Dimitrov, M. & Khadzhimitsev, P. (1961). Normal microflora of Kachkaval cheese and its effect on its composition. *Sborn. Trud. Nauchnoiszl. San.-Khigien. Inst. 1960,* **6**, 137–40 (in Bulgarian); *Dairy Sci. Abstr.* **25** (1963), no. 1925.

Domenichini, G. (1978). Organisms contaminating ewes' milk cheese during ripening. *Sci. Tech. Latterio-Casearia,* **29**, 182–93 (in Italian); *Fd Sci. Technol. Abstr.* **11** (1979), no. 4 P 650.

Dragoni, I., Cantoni, C. & Corti, S. (1983). Mycotoxic contamination of Padano grana cheese by toxigenic specie. *Latte,* **8**, 605–7 (in Italian).

Dubois, G. (1981). Study of *Penicillium* having contaminated yoghurts. *Lait,* **61**, 333–6 (in French).

Duitschaever, C. L. & Irvine, D. M. (1971). A case study: effect of mold on growth of coagulase-positive staphylococci in Cheddar cheese. *J. Milk Fd Technol.,* **34**, 583.

Dulley, J. R. & Houlihan, D. B. (1979). An examination of Queensland milk and cheese for possible aflatoxin contamination. *Aust. J. Dairy Technol.,* **34**, 12–13.

El-Bassiony, T. A., Atia, M. & Khier, F. A. (1980). Search for the predominance of fungi species in cheese. *Assiut Vet Med. J.,* **7**, 175–84; *Fd Sci. Technol. Abstr.* **14** (1982), no. 10 P 1545.

El-Bazza, Z. E., Zedan, H. H., Toama, M. A. & El-Tayeb, O. M. (1983). Isolation of aflatoxin-producing fungi from Egyptian food and feed commodities. In *Proceedings of the International Symposium on Mycotoxins,* Cairo, Egypt, 6–8 September, 1981, ed. K. Naguib, M. M. Naguib, D. L. Park & A. E. Pohland, Food and Drug Administration, Rockville, Maryland, USA and National Research Centre, Cairo, Egypt, pp. 443–54.

El-Essawy, H. A., Saudi, A. M., Mahmoud, S. & Morgan, S. D. (1984). Fungal contamination of hard cheese. *Assiut Vet. Med. J.,* **11**, 125–9; *Fd Sci. Technol. Abstr.* **17** (1985), no. 1 P 208.

El-Refai, A.-M. H., Sallam, L. A. R. & Naim, N. (1970). The alkaloids of fungi. I. The formation of ergoline alkaloids by representative mold fungi. *Jap. J. Microbiol.,* **14**, 91–7.

Engel, G. (1978). Formation of mycotoxins on Tilsit cheese. *Milchwissenschaft,* **33**, 201–3 (in German).

Engel, G. & Prokopek, D. (1980). No detection of patulin and penicill(in)ic acid in cheese produced by *Penicillium roqueforti* — strains forming patulin and penicill(in)ic acid. *Milchwissenschaft,* **35**, 218–20 (in German).

Engel, G. & Teuber, M. (1980). Formation and distribution of sterigmatocystin in cheese after inoculation with *Aspergillus versicolor* and *A. nidulans. Milchwissenschaft,* **35**, 750–6 (in German).

Ferrando, R., Parodi, A., Henry, N., Delort-Laval, J. & N'Diaye, A. L. (1977). 'Milk aflatoxine' et toxicité de relais. *CR Acad. Sci. Paris,* **284**, 855–8.

Fonseca, H., Nogueira, J. N., Graner, M., Oliveira, A. J., Caruso, J. G. B., Boralli, C., Calori, M. A. & Khatounian, C. A. (1983). Natural occurrence of mycotoxins in some Brazilian foods. Part II. In *Research in Food Science and Nutrition, Vol. 3, Human Nutrition,* ed. J. V. McLoughlin & B. M. McKenna. Boole Press Ltd, Dublin, pp. 53–4.

Foster, E. M., Nelson, F. E., Speck, M. L., Doetsch, R. N. & Olson, J. C., Jr (1957). *Dairy Microbiology.* Prentice-Hall, Inc., Englewood Cliffs, New Jersey.

Francis, O. J., Jr, Ware, G. M., Carman, A. S. & Kuan, S. S. (1985). Thin layer chromatographic determination of sterigmatocystin in cheese. *J. Ass. Offic. Anal. Chem.,* **68**, 643–5.

Francis, O. J., Jr, Ware, G. M., Carman, A., Kirschenheuter, G. & Kuan, S. S. (1987). Thin-layer chromatographic determination of sterigmatocystin in cheese: interlaboratory study. *J. Assoc. Off. Anal. Chem.,* **70**, 842–4.

Frank, H. K. (1966). Aflatoxine in Lebensmitteln. *Arch. Lebensmittelhyg.,* **17**, 237–42.

Frank, H. K. (1968). Diffusion of aflatoxin in foodstuffs. *J. Fd Sci.,* **33**, 98–100.

Frank, H. K. & Eyrich, W. (1968). Über den Nachweis von Aflatoxinen und das Vorkmmen Aflatoxin-vortauschender Substanzen in Lebensmitteln. *Z. Lebensmittelunters. u.-Forsch.,* **138**, 1–11.

Frisvad, J. C. (1984). Expressions of secondary metabolism as fundamental characters in *Penicillium* taxonomy. In *Developments in Food Science, Vol. 7,*

Toxigenic Fungi — Their Toxins and Health Hazard, ed. H. Kurata & Y. Ueno. Kodansha, Tokyo and Elsevier, Amsterdam, pp. 98–106.

Frisvad, J. C. (1986). Taxonomic approaches to mycotoxin identification (taxonomic indication of mycotoxin content in foods). In *Modern Methods in the Analysis and Structural Elucidation of Mycotoxins*, ed. R. J. Cole. Academic Press, Inc., Orlando, Florida, pp. 415–57.

Fritz, W. (1983). Investigations on the occurrence of selected mycotoxins in foods. *Z. Ges. Hyg.*, **29**, 650–4 (in German).

Fritz, W. & Engst, R. (1981). Survey of selected mycotoxins in food. *J. Environ. Sci. Health*, **B16**, 193–210.

Fritz, W., Donath, R. & Engst, R. (1977). Determination and occurrence of aflatoxin M_1 and B_1 in milk and dairy products. *Nahrung*, **21**, 79–84 (in German).

Gaddi, B. L. (1973). Mycotoxin-producing potential of fungi isolated from cheese, PhD Thesis. University of Wisconsin, Madison, Wisconsin; *Diss. Abstr. Int. B.* **34** (1974), 5012-B.

Galea, V. & Bara, A. (1976). Determination of aflatoxins in food products. *Igiena*, **25**, 123–8 (in Romanian).

Galli, A. & Zambrini, A. (1978). Surface microflora of Provolone cheese. *Ind. Latte* **14**, 3–12 (in Italian); *Fd Sci. Technol. Abstr.* **10**, no. 11 P 1886.

Galli, A., Zambrini, A. & Süss, L. (1978). Trials with antiparasitic and antifungal agents on Provolone cheese. *Ind. Latte,* **14**, 23–36 (in Italian); *Fd Sci. Techol. Abstr.* **10** (1978), no 11 P 1884.

García, A. M. & Fernańdez, G. S. (1984). Contaminating mycoflora in yogurt: general aspects and special reference to the genus *Penicillium*. *J. Fd Protect.*, **47**, 629–36.

Gareis, M., Reubel, G., Kröning, T. & Porzig, R. (1987). A case of fading puppy syndrome associated with ochratoxin A. *Tierärztl. Umsch.*, **43**, 77–80 (in German).

Gertz, C. & Böschemeyer, L. (1980). A screening method for the determination of various mycotoxins in food. *Z. Lebensmittelunters. u.-Forsch.*, **171**, 335–40 (in German).

Ghosal, S., Biswas, K. & Chakrabarti, D. K. (1979). Toxic naphtho-γ-pyrones from *Aspergillus niger*. *J. Agric. Fd. Chem.*, **27**, 1347–51.

Gilbert, M., Penel, A., Kosikowski, F. V., Henion, J. D., Maylin, G. A. & Lisk, D. J. (1977). Electron affinity gas chromatographic determination of beta-nitropropionic acid as its pentafluorobenzyl derivative in cheeses and mold filtrates. *J. Fd. Sci.*, **42**, 1650–3.

Goto, T., Manabe, M. & Matsuura, S. (1982). Analysis of aflatoxins in milk and milk products by high-performance liquid chromatography. *Agric. Biol. Chem.*, **46**, 801–2.

Gross, M., Levy, R. & Toepke, H. (1984). On the occurrence and analysis of the mycotoxin emodin. *Nahrung*, **28**, 31–44 (in German).

Gueguen, M., Desfleurs, M. & Lemarinier, S. (1978). *Penicillium roqueforti* Thom, responsable d'un nouvel accident en fromagerie de pâtes molles. *Lait,* **58**, 327–35.

Hanssen, E. & Jung, M. (1973). Control of aflatoxins in the food industry. *Pure Appl. Chem.* **35**, 239–50.

Harwig, J., Kuiper-Goodman, T. & Scott, P. M. (1983). Microbial food toxicants:

ochratoxins. In *Handbook of Foodborne Diseases of Biological Origin*, ed. M. Rechcigl Jr. CRC Press, Boca Raton, Florida, pp. 193–238.

Hendricks, J. D., Sinnhuber, R. O., Wales, J. H., Stack, M. E. & Hsieh, D. P. H. (1980). Hepatocarcinogenicity of sterigmatocystin and versicolorin A to rainbow trout (*Salmo gairdneri*) embryos. *J. Natn. Cancer Inst.*, **64**, 1503–9.

Ichinoe, M. & Kurata, H. (1983). Trichothecene producing fungi. In *Developments in Food Science, Vol. 4, Trichothecenes — Chemical, Biological and Toxicological Aspects*, ed. Y. Ueno. Kodansha, Tokyo and Elsevier, Amsterdam, pp. 73–82.

Inagaki, N. (1962). On some fungi isolated from foods (I). *Trans. Mycol. Soc. Japan* **4**, 1–5.

Iwasaki, T. & Kosikowski, F. V. (1973). Production of β-nitropropionic acid in foods. *J. Fd Sci.*, **38**, 1162–5.

Iwata, K., Nagai, T. & Okudaira, M. (1969). Fumigatoxin, a new toxin from a strain of *Aspergillus fumigatus*. *J. S. Afr. Chem. Inst.*, **22**, S131–S141.

Jacobson, W. C., Harmeyer, W. C. & Wiseman, H. G. (1971). Determination of aflatoxins B_1 and M_1 in milk. *J. Dairy Sci.*, **54**, 21–4.

Jacquet, J. & Boutibonnes, P. (1970). Researches on flavacoumarins (flavatoxins or aflatoxins). *Rev. Immunol., Paris*, **34**, 245–74 (in French).

Jacquet, J. & Desfleurs, M. (1966a). *Scopulariopsis brevicaulis* Bainier ou mieux *Scopulariopsis fusca* Zach., agent de taches superficielles brun violet des fromages à pâte molle. *Lait*, **46**, 241–53.

Jacquet, J. & Desfleurs, M. (1966b). *Cladosporium herbarum* Link, agent d'accidents tardifs de 'bleu' sur les fromages à pâte molle, et spécialement le camembert. *Lait*, **46**, 485–97.

Jacquet, J. & Lafont, P. (1979). Sur la présence de mycotoxines dans les aliments. *CR Acad. Agric. Fr.*, **65**, 1519–27.

Jacquet, J. & Tantaoui-Elaraki, A. (1976). Les produits laitiers comme mileux de culture et de toxicogénèse des *Aspergillus* du groupe *flavus*. Cas particulier des fromages. *CR Acad. Agric. Fr.*, **62**, 208–17.

Jacquet, J. & Teherani, M. (1974). Présence exceptionnelle de l'aflatoxine dans certains produits d'origine animale. Rôle possible du poivre. *Bull. Acad. Vét. Fr.*, **47**, 313–15.

Jacquet, J., Boutibonnes, P. & Teherani, A. (1970). Sur la présence des flavatoxines dans les aliments des animaux et dans les aliments d'origine animale destinés à l'homme. *Bull. Acad. Vét.*, **43**, 35–43.

Jantea, F., Oprişescu, D., Stǎncescu, B., Molnar, A. & Sepeţeanu, A. (1972). Hygienic significance of fungi in foods. *Igiena*, **21**, 495–504 (in Romanian); *Dairy Sci. Abstr.*, **35** (1973), no. 2264.

Jarvis, B. (1972). Mould spoilage of foods. *Process Biochem.*, **7**, 11–13.

Jarvis, B. (1983). Mould and mycotoxins in mouldy cheese. *Microbiol. Alim. Nutr.*, **1**, 187–191.

Jarvis, B., Williams, A. P., Hocking, A. D., Chapman, W. B., Cooper, S. J., Wood, G. M. & Pitt, J. I. (1985). Occurrence and significance of penicillia and other fungi in mould-spoiled foods. In *Trichothecenes and Other Mycotoxins*, ed. J. Lacey, John Wiley & Sons Ltd, Chichester, pp. 193–205.

Jesenská, Z. & Poláková, O. (1978). Problems of the presence of potential mycotoxin producers in milk powders for babies. *Z. Lebensmittelunters Forsch.*, **166**, 1–4 (in German).

Jizdny, J. (1982). Method for determining ochratoxin A in raw materials and feed mixtures. *Krmivarstvi Sluzby,* **19**, 94–6 (in Czech); *Chem. Abstr.,* **99** (1983), 37165c.

Josefsson, E. (1981). Mycotoxins in cheese. *Vår Föda,* **33**, 237–48 (in Swedish).

Jully, B. W. (1977). Control of moulds by a surface treatment of cheeses. *Latte,* **2**, 413–15 (in Italian); *Fd Sci. Tech. Abstr.* **11** (1979), no. 7 P 1228.

Kamimura, H., Nishijima, M., Yasuda, K., Ushiyama, H., Tabata, S., Matsumoto, S. & Nishima, T. (1985). Simple, rapid cleanup method for analysis of aflatoxins and comparison with various methods. *J. Ass. Offic. Anal. Chem.,* **68**, 458–61.

Karaioannoglou, P. (1984). Aflatoxin production on white brined Feta cheese. *Milchwissenschaft,* **39**, 671–4.

Keilling, J., Casalis, J., Duthen, J., Sigonney, L. & Glaser, I. (1947). Sur l'accident du 'Bleu' en fromagerie de pâtes molles à croûte fleurie. *Lait,* **27**, 461–6.

Keilling, J., Casalis, J., Souignac, G. & Dubreuil, G. (1956). Sur une altération superficielle des fromages à pâte molle et croûte fleurie provoquée par *P. funiculosum. Lait,* **36**, 241–50.

Kiermeier, F. (1970). Analysis of aflatoxins in cheese. *Z. Lebensmittelunters. u.-Forsch.,* **144**, 293–7 (in German).

Kiermeier, F. (1971*a*). Zur Aflatoxin-Bildung auf Käse. *Ernahrungsforschung,* **16**, 519–26.

Kiermeier, F. (1971*b*). On aflatoxin formation in milk and milk products. IV. Model experiments with milk powder. *Z. Lebensmittelunters. u.-Forsch.,* **146**, 262–5 (in German).

Kiermeier, F. (1981). Mykotoxine in Milch und Milchprodukten. In *Mykotoxine in Lebensmitteln,* ed. J. Reiss. Gustav Fischer Verlag, Stuttgart, pp. 245–71.

Kiermeier, F. & Behringer, G. (1972). On aflatoxin formation in milk and milk products. VII. Difficulties in sampling of foods containing aflatoxin. *Z. Lebensmittelunters. u.-Forsch.,* **148**, 72–6 (in German).

Kiermeier, F. & Behringer, G. (1973). Influence of pH on aflatoxin formation in model experiments with milk powder. *Z. Lebensmittelunters. u.-Forsch.,* **151**, 392–4 (in German).

Kiermeier, F. & Behringer, G. (1977*a*). Influence of cooling temperatures on aflatoxin formation in milk products. *Z. Lebensmittelunters. u.-Forsch.,* **164**, 283–5 (in German).

Kiermeier, F. & Behringer, G. (1977*b*). Sampling of cheese for aflatoxins. *Z. Lebensmittelunters. u.-Forsch.,* **165**, 30–3 (in German).

Kiermeier, F. & Böhm, S. (1971). On aflatoxin formation in milk and milk products. V. Application of the hen-embryo-test for the affirmation of thin layer chromatographic determination of aflatoxin in cheese. *Z. Lebensmittelunters. u.-Forsch.,* **147**, 61–4 (in German).

Kiermeier, F. & Gross, D. (1970*a*). Determination of aflatoxin B_1 in cheese. *Z. Lebensmittelunters. u.-Forsch.,* **142**, 120–3 (in German).

Kiermeier, F. & Groll, D. (1970*b*). About the formation of aflatoxin B_1 in cheese. *Z. Lebensmittelunters. u.-Forsch.,* **143**, 81–9 (in German).

Kiermeier, F. & Kraus, P.-V. (1980). On the possible presence of sterigmatocystin in milk and its behaviour in cheese. *Z. Lebensmittelunters. u.-Forsch.,* **170**, 421–4 (in German).

Kiermeier, F. & Rumpf, S. (1975). Behaviour of aflatoxin during production of processed cheese. Z. Lebensmittelunters. u.-Forsch., **157**, 211–16 (in German).

Kiermeier, F. & Weiss, G. (1976). Investigations of aflatoxins B_1, B_2, G_1, G_2, and M_1 in milk and milk products. Z. Lebensmittelunters. u.-Forsch., **160**, 337–44 (in German).

Kiermeier, F. & Zierer, E. (1975). Effect of pimaricin on moulds and their aflatoxin formation in cheese. Z. Lebensmittelunters. u.-Forsch., **157**, 253–62 (in German).

King, A. D., Jr. & Schade, J. E. (1984). Alternaria toxins and their importance in food. J. Fd Protect., **47**, 886–901.

Koch, W. & Kross, W. (1986). Article on the quantitative determination of harmful aflatoxins in selected samples of cheese of the military subsistence. Wehrmed. Monatschr., **30**, 203–6 (in German).

Korzybski, T., Kowszyk-Gindifer, Z. & Kuryłowicz, W. (1967). Antibiotics. Origin, Nature and Properties, Vol. II. Pergamon Press, Oxford.

Kozareva, M., Shalamanova, V., Stefanova, M. & Ilieva, K. (1972). Microbiological characteristics of fresh butter. Khig. Zdraveopazvane, **15**, 172–77 (in Bulgarian); Dairy Sci. Abstr. **35** (1973), no. 2260.

Lafont, P., Siriwardana, M. G., Combemale, I. & Lafont, J. (1979a). Mycophenolic acid in marketed cheeses. Fd Cosmet. Toxicol., **17**, 147–9.

Lafont, P., Siriwardana, M. G. & Lafont, J. (1979b). Contamination de fromages par des métabolites fongiques. Méd Nutr., **15**, 257–62.

Landmark, E. & Aukrust, L. (1985). High-performance liquid chromatography of Cladosporium herbarum. Identification of allergens with immunological techniques. Int. Arch. Allergy Appl. Immunol., **78**, 71–6.

Laub, E. & Woller, R. (1977). Vorkommen der Aflatoxine B_1, B_2, G_1 und G_2 in Lebensmittelproben des Handels. Dtsche Lebensmittel-Rundsch., **73**, 8–10.

Le Bars, J. (1985). Fungal contamination in milk replacers for calves: nature and sources. Sci. Aliments, **5**, 157–62 (in French).

Leistner, L. & Eckardt, C. (1979). Occurrence of toxinogenic Penicillia in meat products. Fleischwirtschaft, **59**, 1892–6 (in German).

Leistner, L. & Pitt, J. I. (1977). Miscellaneous Penicillium toxins. In Mycotoxins in Human and Animal Health, ed. J. V. Rodricks, C. W. Hesseltine & M. A. Mehlman. Pathotox, Park Forest South, Illinois, pp. 639–53.

Lembhe, A. F., Ranganathan, B., Rao, M. V. R. & Rao, L. K. (1981). Aflatoxin production in khoa. J. Fd Protect., **44**, 137–8, 143.

Lie, J. L. & Marth, E. H. (1967). Formation of aflatoxin in Cheddar cheese by Aspergillus flavus and Aspergillus parasiticus. J. Dairy Sci., **50**, 1708–10.

Lie, J. L. & Marth, E. H. (1968). Aflatoxin formation by Aspergillus flavus and Aspergillus parasiticus in a casein substrate at different pH values. J. Dairy Sci., **51**, 1743–7.

Lieu, F. Y. & Bullerman, L. B. (1977). Production and stability of aflatoxins, penicillic acid and patulin in several substrates. J. Fd Sci., **42**, 1222–4, 1228.

Lück, H. & Cheesman, C. E. (1978). Mould growth on cheese as influenced by pimaricin (natamycin) or sorbate treatments. S. Afr. J. Dairy Technol., **10**, 143–6.

Lück, H. & Wehner, F. C. (1979). Mycotoxins in milk and dairy products. S. Afr. J. Dairy Technol., **11**, 169–76.

Lück, H., Steyn, M. & Wehner, F. C. (1976*a*). A survey of milk powder for aflatoxin content. *S. Afr. J. Dairy Technol.,* **8**, 85–6.

Lück, H., Wehner, F. C., Plomp, A. & Steyn, M. (1976*b*). Mycotoxins in South African cheeses. *S. Afr. J. Dairy Technol.,* **8**, 107–10.

Mahmoud, S. M., Abd-El Rahman, H. A., Morgan, S. D. & Hafez, R. S. (1983). Mycological studies on Egyptian soft cheese and cooking butter. *Assiut Vet. Med. J.,* **11**, 151–5; *Fd Sci. Technol. Abstr.* **16** (1984), no. 7 P 1497.

Mahrous, A. M., Naguib, M. M., Naguib, K., Pohland, A. E. & Campbell, A. D. (1978). Aflatoxin residues in Egyptian foods. II. Milk and dairy products. *Abstr. XII Intern. Congr. Microbiology* (1978); cited by Pfleger (1985).

Malik, R. K., Engel, G. & Teuber, M. (1986). Effect of some nutrients on the production of cyclopiazonic acid by *Penicillium verrucosum* var. *cyclopium. Appl. Microbiol. Biotechnol.,* **24**, 71–4.

Marth, E. H. (1978). Dairy products. In *Food and Beverage Mycology*, ed. L. R. Beuchat. Avi Publishing Co., Inc., Westport, Connecticut, pp. 145–72.

Masaly, R. I., El-Deeb, S. A., Ismail, A. A. & Yousef, A. (1981). Contamination of dairy products and cattle feed concentrates on the local market with aflatoxin. *Egypt. J. Dairy Sci.,* **9**, 181–91; *Chem. Abstr.* **96** (1982), 84238b.

Masri, M. S., Garcia, V. C. & Page, J. R. (1969). The aflatoxin M content of milk from cows fed known amounts of aflatoxin. *Vet. Rec.,* **84**, 146–7.

Matsumoto, H., Unrau, A. M., Hylin, J. W. & Temple, B. (1961). Spectrophotometric determination of 3-nitropropanoic acid in biological extracts. *Anal. Chem.,* **33**, 1442–4.

Mattsson, N. (1977). Mould control on cheese. Experience with potassium sorbate (added to brine) and pimaricin for surface treatment of hard cheese. *Svenska Mejeritidningen,* **69**, 14–15 (in Swedish); *Fd Sci. Technol. Abstr.* **10** (1978), no. 7 P 1057.

Megalla, S. E. & Hafez, A. H. (1982). Detoxification of aflatoxin B_1 by acidogenous yoghurt. *Mycopathologia,* **77**, 89–91.

Megalla, S. E. & Mohran, M. A. (1984). Fate of aflatoxin B-1 in fermented dairy products. *Mycopathologia,* **88**, 27–9.

Megalla, S. E., Kamel, Y. Y., Abdel-Fattah, H. M. & Hafez, A. H. (1981). Determining the kinetic behaviour for the secretion of milk toxin as related to dosage level of aflatoxin B_1. *Z. Ernahrungswiss.,* **20**, 216–22.

Mehran, M., Behboodi, M. & Rhoubakhsh-Kh, A. (1975). Microbial contaminations of Iranian white cheese produced from raw milk. *J. Dairy Sci.,* **58**, 784.

Metwally, M. M., Dawood, A. E. A. & Aly, A.-E. N. (1983). Application of the minicolumn technique for screening dairy and selected food products for aflatoxin B_1 and G_1. In *Proceedings of the International Symposium on Mycotoxins*, Cairo, Egypt, 6–8 September 1981, ed. K. Naguib, M. M. Naguib, D. L. Park & A. E. Pohland. Food and Drug Administration, Rockville, Maryland, USA and National Research Centre, Cairo, Egypt, pp. 455–8.

Mihai, M. D., Turburi, A. & Taga, M. (1970). The presence of aflatoxigenic fungi in cheese and their capacity to synthetize aflatoxins. *Archiva Vet.,* **6**, 113–18.

Mislivec, P. B. (1985). Mycotoxins of the Aspergilli. In *Aspergillosis*, ed Y. Aldoory & G. E. Wagner. Charles C. Thomas, Springfield, Illinois, pp. 257–68.

Mislivec, P. B., Hunter, J. H. & Tuite, J. (1968). Assay for aflatoxin production by the genera *Aspergillus* and *Penicillium*. *Appl. Microbiol.*, **16**, 1053–5.

Moreau, C. (1976*a*). Mycotoxins in dairy products. *Lait*, **56**, 286–303 (in French).

Moreau, C. (1983). Some problems of moulds in the dairy industry. *Techn. Laitière*, no. 975, 37–8, 41–3 (in French); *Fd Sci. Technol. Abstr.*, **16** (1984), no. 9 P 1874.

Moreau, M. (1976*b*). Pollution ambiante et accidents de fabrication par moisissures des fromages à pâtes molles. *Ind. Alim. Agric.*, **93**, 1315–29.

Morris, H. A. & Castberg, H. B. (1980). Control of surface growth on Blue cheese using pimaricin. *Cult. Dairy Prod. J.*, **15**, 21–3; *Fd Sci. Technol. Abstr.*, **12** (1980), no. 12 P 2067.

Moss, M. O. (1977). Aspergillus mycotoxins. In *Genetics and Physiology of Aspergillus, Vol. 1*, ed. J. E. Smith. Academic Press, London, pp. 499–524.

Moubasher, A. H., Abdel-Kader, M. I. A. & El-Kady, I. A. (1978). Toxigenic fungi isolated from Roquefort cheese. *Mycopathologia*, **66**, 187–90.

Muys, G. T., Van Gils, H. W. & de Vogel, P. (1966). The determination and enumeration of the associative microflora of edible emulsions. *Lab. Practice*, **15**, 975–84.

Nabney, J., Burbage, M. B., Allcroft, R. & Lewis, G. (1967). Metabolism of aflatoxin in sheep: excretion pattern in the lactating ewe. *Fd Cosmet. Toxicol.*, **5**, 11–17.

Naguib, K., Naguib, M. M., Monib, A., Nour, M. A., El-Khadem, M. & Hosny, I. M. (1983). Studies on mycotoxins in local Roquefort cheese. In *Proceedings of the International Symposium on Mycotoxins*, Cairo, Egypt, 6–8 September 1981, ed. K. Naguib, M. M. Naguib, D. L. Park & A. E. Pohland. Food and Drug Administration, Rockville, Maryland, USA and National Research Centre, Cairo, Egypt, pp. 301–4.

Nakae, T. & Yoneya, T. (1976). Studies on the fungi contaminating milk and milking environment. II. The influence of the fungi isolated on the quality of milk and the fungal toxicity bioassay by chick embryo test. *Jap. J. Dairy Fd Sci.*, **25**, A113–A118 (in Japanese); *Fd Sci. Technol. Abstr.*, **9** (1977), no. 3 P 539.

Nilson, K. M., Shahani, K. M., Vakil, J. R. & Kilara, A. (1975). Pimaricin and mycostatin for retarding Cottage cheese spoilage. *J. Dairy Sci.*, **58**, 668–71.

Northolt, M. D. & de Boer, E. (1981). Moulds and yeasts in powdered cheese. *Warenchemicus*, **11**, 112–15 (in Dutch); *Fd Sci. Technol. Abstr.*, **16** (1984), no. 3 P 604.

Northolt, M. D. & Van Egmond, H. P. (1982*a*). Contamination of ripening cheese with *Aspergillus versicolor* and sterigmatocystin. In *Proceedings of a Fourth Meeting on Mycotoxins in Animal Disease*, 1–3 April 1981, Weybridge, UK, ed. G. A. Pepin, D. S. P. Patterson & D. E. Gray. Ministry of Agriculture, Fisheries and Food, Alnwick, Northumberland, UK, pp. 90–2.

Northolt, M. D. &. Van Egmond, H. P. (1982*b*). Limits of water activity and temperature for the production of some mycotoxins. In *Proceedings of a Fourth Meeting on Mycotoxins in Animal Disease*, 1–3 April 1981, Weybridge, UK, ed. G. A. Pepin, D. S. P. Patterson & D. E. Gray. Ministry of Agriculture, Fisheries and Food, Alnwick, Northumberland, UK, pp. 106–8.

Northolt, M. D. & Soentoro, P. S. S. (1979). Fungal species on Dutch cheese. *Neth. Milk Dairy J.,* **33**, 205–8.

Northolt, M. D., Van Egmond, H. P. & Paulsch, W. E. (1978). Patulin production by some fungal species in relation to water activity and temperature. *J. Fd Protect.,* **41**, 885–90.

Northolt, M. D., Van Egmond, H. P. & Paulsch, W. E. (1979*a*). Penicillic acid production by some fungal species in relation to water activity and temperature. *J. Fd Protect.,* **42**, 476–84.

Northolt, M. D., Van Egmond, H. P. & Paulsch, W. E. (1979*b*). Ochratoxin A production by some fungal species in relation to water activity and temperature. *J. Fd Protect.,* **42**, 485–90.

Northolt, M. D., Van Egmond, H. P., Soentoro, P. & Deijll, E. (1980). Fungal growth and the presence of sterigmatocystin in hard cheese. *J. Ass. Offic. Anal. Chem.,* **63**, 115–19.

Nowotny, P., Baltes, W., Krönert, W. & Weber, R. (1983*a*). Untersuchung von Käseproben des Handels auf die Mykotoxine Sterigmatocystin, Citrinin und Ochratoxin A. *Lebensmittelchem. Gerichtl. Chem.,* **37**, 71–2.

Nowotny, P., Baltes, W., Krönert, W. & Weber, R. (1983*b*). Thin layer chromatographic method for the determination of 22 mycotoxins in mouldy food. *Chem. Mikrobiol. Technol. Lebensm.,* **8**, 24–8 (in German).

Nuñez, M., Medina, M., Gaya, P. & Dias-Amado, C. (1981). The yeasts and moulds of Spanish mould-ripened Cabrales cheese. *Lait,* **61**, 62–79 (in French).

Oldham, L. S., Oehme, F. W. & Kelley, D. C. (1971). Production of aflatoxin in pre-packaged luncheon meat and cheese at refrigerator temperatures. *J. Milk Fd Technol.,* **34**, 349–51.

Olivigni, F. J. & Bullerman, L. B. (1977). Simultaneous production of penicillic acid and patulin by a *Penicillium* species isolated from Cheddar cheese. *J. Fd Sci.,* **42**, 1654–7, 1665.

Olivigni, F. J. & Bullerman, L. B. (1978). Production of penicillic acid and patulin by an atypical *Penicillium roqueforti* isolate. *Appl. Environ. Microbiol.,* **35**, 435–8.

Olšanský, Č., Vokáčová, H. & Stryková, M. (1979). Recontamination of some dairy products with yeasts and moulds. *Průmysl Potravin,* **30**, 96–8 (in Slovakian); *Fd Sci. Technol. Abstr.* **11** (1979), no. 11 P 1850.

Olsen, N. F. (1979). Moldy cheese — what should consumers do with it? *Dairy Ice Cream Field,* **162**, 82C, 82E.

Park, K.-Y. & Bullerman, L. B. (1983). Effects of substrate and temperature on aflatoxin production by *Aspergillus parasiticus* and *Aspergillus flavus. J. Fd Protect.,* **46**, 178–84.

Patterson, M. F. & Damoglou, A. P. (1985). Identification and potential toxicity of fungi isolated from mould-spoiled foods. *Rec. Agric. Res.,* **33**, 49–55.

Paul, R., Kalra, M. S. & Singh, A. (1976). Incidence of aflatoxins in milk and milk products. *Indian J. Dairy Sci.,* **29**, 318–21.

Pérez, B. S., Asenjo, P. P., Iguácel, J. T. & Lorenzo, P. L. (1973). Investigacion de aflatoxinas en diversos alimentos españoles. *An. Bromatol.,* **25**, 297–319 (in Spanish).

Pfleger, R. (1985). Mykotoxine in Käse — alte und neue Fragen. *Milchwirtsch. Ber. Bundesanstalt Wolfpassing Rotholz,* **85**, 297–301.

Piskorska-Pliszczyńska, J. & Borkowska-Opacka, B. (1984). Natural occurrence of ochratoxin A and two ochratoxin-producing fungal strains in cheesecake. *Bull. Vet. Inst. Pulawy,* **27**, 95–8.

Pitt, J. I. (1979*a*). *Penicillium crustosum* and *P. simplicissimum,* the correct names for two common species producing tremorgenic mycotoxins. *Mycologia,* **71**, 1166–77.

Pitt, J. I. (1979*b*). The genus *Penicillium* and its teleomorphic states *Eupenicillium* and *Talaromyces.* Academic Press, London.

Pitt, J. I. & Hocking, A. D. (1985). *Fungi and Food Spoilage.* Academic Press, Sydney, Australia.

Podojil, M., Sedmera, P., Vokoun, J., Betina, V., Baráthová, H., Ďuračková, Z., Horákova, K. & Nemec, P. (1979). *Eurotium (Aspergillus) repens* metabolites and their biological activity. *Folia Microbiol.,* **23**, 438–43.

Pohland, A. E. & Allen, R. (1970). Analysis and chemical confirmation of patulin in grains. *J. Ass. Offic. Anal. Chem.,* **53**, 686–7.

Pohlmeier, M. M. & Bullerman, L. B. (1978). Ochratoxin production by a *Penicillium* species isolated from cheese. *J. Fd Protect.,* **41**, 829.

Polonelli, L., Orsini, D. & Morace, G. (1984). Toxin producing potential of fungi isolated from food. *Igiene Moderna,* **81**, 483–8; *Fd Sci. Technol. Abstr.,* **18** (1986), no. 7 C 1.

Polzhofer, K.-P. (1977*a*). Simulation of natural mould growth in processed cheese. *Z. Lebensmittelunters. u.-Forsch.,* **164**, 94–5 (in German).

Polzhofer, K.-P. (1977*b*). Determination of aflatoxins in milk and milk products. *Z. Lebensmittelunters. u.-Forsch.,* **163**, 175–7 (in German).

Rabie, C. J. (1986). Important lesser known toxigenic fungi. In *Bioactive Molecules, Vol. 1, Mycotoxins and Phycotoxins,* ed. P. S. Steyn & R. Vleggaar. Elsevier Science Publishers, Amsterdam, p. 29–40.

Ray, L. L. & Bullerman, L. B. (1982). Preventing growth of potentially toxic molds using antifungal agents. *J. Fd Protect.,* **45**, 953–63.

Recordon, J., Hardy, J. & Wéber, F. (1980). Enquête sur les accidents dus au développement de moisissures de la famille des Mucorales en fromagerie de pâtes molles. Rapport de synthèse et conclusions. *Bull. École Nat. Sup. Agron. Ind. Alim.,* **22**, 9–21.

Rehm, H.-J. & Schmidt, I. (1969). Production of aflatoxins in butter and margarine. *Z. Lebensmittelunters. u.-Forsch.,* **140**, 164–5 (in German).

Richard, J. L. & Arp, L. H. (1979). Natural occurrence of the mycotoxin penitrem A in moldy cream cheese. *Mycopathologia,* **67**, 107–9.

Roberts, B. A. & Patterson, D. S. P. (1975). Detection of twelve mycotoxins in mixed animal feedstuffs, using a novel membrane cleanup procedure. *J. Ass. Offic. Anal. Chem.,* **58**, 1178–81.

Rothenbühler, E. & Bachmann, M. (1970). Formation of aflatoxin in cheese. *Schweiz. Milchzeitung (Wissenschaftliche Beilage Nr. 123),* **96**, 1053–6 (in German); *Dairy Sci. Abstr.* **33** (1971), no. 1009.

Saito, K., Nishijima, M., Kamimura, H., Ibe, A., Ochiai, S. & Naoi, Y. (1979). Analytical method and investigation of natural occurrence of aflatoxins in dairy products (Studies on mycotoxins in foods. IX). *J. Fd Hyg. Soc. Jap.,* **20**, 27–32 (in Japanese).

Sasaki, Y. (1950). A study of molds in butter. *J. Fac. Agric. Hokkaido Univ.*, **49**, 121–249 (in Japanese); cited by Moreau (1976a).

Schneider, D. J., Marasas, W. F. O., Collett, M. G. & van der Westhuisen, G. C. A. (1985). An experimental mycotoxicosis in sheep and goats caused by *Drechslera campanulata*, a fungal pathogen of green oats. *Onderstepoort J. Vet. Res.*, **52**, 93–100.

Scott, P. M. (1977). *Penicillium* mycotoxins. In *Mycotoxic Fungi, Mycotoxins, Mycotoxicoses. An Encyclopedic Handbook, Vol. 1*, ed. T. D. Wyllie & L. C. Morehouse. Marcel Dekker, New York, pp. 283–356.

Shih, C. N. & Marth, E. H. (1969). Aflatoxins not recovered from commercial mold-ripened cheeses. *J. Dairy Sci.*, **52**, 1681–2.

Shih, C. N. & Marth, E. H. (1971). A procedure for rapid recovery of aflatoxins from cheese and other foods. *J. Milk Fd Technol.*, **34**, 119–23.

Shih, C. N. & Marth, E. H. (1972a). Experimental production of aflatoxin on brick cheese. *J. Milk Fd Technol.*, **35**, 585–7.

Shih, C. N. & Marth, E. H. (1972b). Production of aflatoxin in a medium fortified with sodium chloride. *J. Dairy Sci.*, **55**, 1415–19.

Shotwell, O. L. & Ellis, J. J. (1976). *Helminthosporium, Drechslera* and *Bipolaris* toxins. In *Mycotoxins and Fungal Related Food Problems. Adv. Chem Ser. no. 149*, ed. J. V. Rodricks. American Chemical Society, Washington, DC, pp. 318–43.

Simón, M. T. C. de & Fernandez, G. S. (1983). Study of the mycoflora present in milky flour for children's consumption [infants]. *An. Bromatol.*, **35**, 79–86 (in Spanish); *Fd Sci. Technol. Abstr.*, **17** (1985), no. 1 G 16.

Siriwardana, M. G. & Lafont, P. (1979). Determination of mycophenolic acid, penicillic acid, patulin, sterigmatocystin, and aflatoxins in cheese. *J. Dairy Sci.*, **62**, 1145–8.

Škrinjar, M. (1985). *Penicillium verrucosum* Dierckx var. *cyclopium* (Westling) Samson, Stolk & Hadlok: appearance in Edam cheese, feasibility of ochratoxin A production and toxicity. *Prehambeno-tehnol. Rev.*, **23**, 35–8 (in Serbian).

Škrinjar, M. & Žakula, R. (1985). Mycotoxins in Edam cheese and their toxicity. *Mljekarstvo*, **35**, 131–7 (in Serbian); *Chem. Abstr.*, **103** (1985), 213569z.

Škrinjar, M., Žakula, R. & Stojanovic, E. (1983). Isolation and determination of moulds in raw milk. *Mljekarstvo*, **33**, 227–30 (in Serbian); *Fd Sci. Technol. Abstr.*, **16** (1984), no. 11 P 2398.

Sozzi, T. & Shepherd, D. (1971). The factors influencing the development of undesirable moulds during the maturation of cheese. *Milchwissenschaft*, **26**, 280–2.

Spillmann, H. & Geiges, O. (1983). Identification of yeasts and moulds in inflated [blown] yoghurts. *Milchwissenschaft*, **38**, 129–32 (in German); *Fd Sci. Technol. Abstr.* **16** (1984), no. 5 P 1100.

Steering Group on Food Surveillance. (1987). *Mycotoxins*. Her Majesty's Stationery Office, London.

Steyn, P. S. (1981). Multimycotoxin analysis. *Pure Appl. Chem.*, **53**, 891–902.

Stoloff, L. (1982). Report on mycotoxins. *J. Ass. Offic. Anal. Chem.*, **65**, 316–23.

Stott, W. T. & Bullerman, L. B. (1976). Instability of patulin in Cheddar cheese. *J. Fd Sci.*, **41**, 201–3.

Šutić, M., Mitić, S. & Svilar, N. (1979). Aflatoxins in milk and milk products.

Mljekarstvo, **29**, 74–80 (in Serbian); *Chem. Abstr.,* **91** (1979), 191557y; *Fd Sci. Technol. Abstr.,* **11** (1979), no. 12 P 2107.

Suzangar, M., Emami, A. & Barnett, R. (1976). Aflatoxin contamination of village milk in Isfahan. Iran, *Trop. Sci.,* **18**, 155–9.

Takahashi, C., Sekita, S., Yoshihira, K. & Natori, S. (1976). The structures of toxic metabolites of *Aspergillus candidus.* II. The compound B (xanthoascin), a hepato- and cardio-toxic xanthocillin analog. *Chem. Pharm. Bull.,* **24**, 2317–21.

Tantaoui-Elaraki, A. & Khabbazi, N. (1984). Possible contamination of cheese by mycotoxins: a review. *Lait,* **64**, 46–71 (in French).

Terao, K. (1983). Sterigmatocystin — a masked potent carcinogenic mycotoxin. *J. Toxicol. — Toxin Reviews,* **2**, 77–110.

Texdorf, I. (1972). Untersuchungen über das Vorkommen von Schimmelpilzen in Anlieferungsmilch. *Arch. Lebensmittelhyg.,* **23**, 99–100.

Thomas, S. B. & Druce, R. G. (1971). Psychrotropic micro-organisms in butter. A review. Part 2. *Dairy Ind.,* **36**, 145–50.

Torrey, G. S. & Marth, E. H. (1977*a*). Isolation and toxicity of molds from foods stored in homes. *J. Fd Protect.,* **40**, 187–90.

Torrey, G. S. & Marth, E. H. (1977*b*). Temperatures in home refrigerators and mold growth at refrigeration temperatures. *J. Fd Protect.,* **40**, 393–7.

Toya, N. (1985). Detection of aflatoxins in milk and milk products. *Bull, Kumamoto Women's University* **37**, 104–10 (in Japanese).

Tsai, W. Y. J. & Bullerman, L. B. (1985). Toxicity of molds isolated from moldy surplus commodity cheeses. *J. Fd Protect.,* **48**, 910.

Ueno, Y. (1985). The toxicology of mycotoxins. *CRC Crit. Rev. Toxicol.,* **14**, 99–132.

Van Egmond, H. P. & Paulsch, W. E. (1986). Mycotoxins in milk and milk products. *Neth. Milk Dairy J.,* **40**, 175–88.

Van Egmond, H. P., Paulsch, W. E., Deijll, E. & Schuller, P. L. (1980). Thin layer chromatographic method for analysis and chemical confirmation of sterigmatocystin in cheese. *J. Ass. Offic. Anal. Chem.,* **63**, 110–14.

Van Egmond, H. P., Deyll, W. E. & Paulsch, W. E. (1982*a*). Analytical method 2-thin layer chromatographic determination of sterigmatocystin in cheese. In *Environmental Carcinogens — Selected Methods of Analysis. Volume 5 — Some Mycotoxins,* ed. H. Egan, L. Stoloff, M. Castegnaro, P. Scott, I. K. O'Neill, H. Bartsch & W. Davis. International Agency for Research on Cancer, Lyon, France, pp. 303–10.

Van Egmond, H. P., Paulsch, W. E. & Northolt, M. D. (1982*b*). Distribution and stability of stergmatocystin in hard cheese. In *Proceedings of a Fourth Meeting on Mycotoxins in Animal Disease,* 1–3 April 1981, Weybridge, UK, ed. G. A. Pepin, D. S. P. Patterson & D. E. Gray. Ministry of Agriculture, Fisheries and Food, Alnwick, Northumberland, UK, pp. 87–9.

Van Egmond, H. P., Paulsch, W. E. & Sizoo, E. A. (1982*c*). Screening procedure for the detection and confirmation of identity of sterigmatocystin in hard cheese. In *Proceedings of a Fourth Meeting on Mycotoxins in Animal Disease,* 1–3 April 1981, Weybridge, UK, ed. G. A. Pepin, D. S. P. Patterson & D. E. Gray. Ministry of Agriculture, Fisheries and Food, Alnwick, Northumberland, UK, pp. 60–2.

Van Walbeek, W., Scott, P. M. & Thatcher, F. S. (1968). Mycotoxins from food-borne fungi. *Can. J. Microbiol.,* **14**, 131–7.

Vanderhoven, C., Remacle, J. & Ramaut, J. L. (1970). Recherche d'un rapport éventuel entre la morphologie de diverses souches d'*Aspergillus flavus* Link et leur production d'aflatoxines. *Rev. Ferment. Ind. Alim.,* **25**, 179–183.

Verdian, N. M. (1981). Contamination of cheese by fungi. *Mikol. Fitopatol,* **15**, 161–4 (in Russian).

Veselý, D., Veselá, D., Kusák, V. & Nesnídal, P. (1978). Distribution of perorally administered aflatoxin B_1 in the tissues and organs of goat (Capra). *Vet. Med. (Prague),* **23**, 555–8 (in Czech).

Veselý, D., Veselá, D. & Jelínek, R. (1982). Nineteen mycotoxins tested on chicken embryos. *Toxicol. Lett.,* **13**, 239–45.

Veselý, D., Veselá, D. & Jelínek, R. (1984). Use of chick embryo in screening for toxin-producing fungi. *Mycopathologia,* **88**, 135–40.

Vijay, H. M., Tsang, P., Young, N. M., Copeland, D. F. & Bernstein, I. L. (1985). Isolation of allergens from *Cladosporium cladosporioides*. *J. Allergy Clin. Immunol.,* **75**, 117.

Villarejo, M. J., Cosano, G. Z., Salinas, R. J., Villar, L. M. P. & Lora, R. P. (1984). Investigation of aflatoxins in natural, sterilized and powdered milk. *Arch. Zootec.,* **33**, 189–98 (in Spanish).

Wallhäuser, K. H. & Lück, E. (1970). Der Einfluss der Sorbinsäure auf mycotoxinbildende Pilze in Lebensmitteln. *Dtsche Lebensmittel-Rundsch.,* **66**, 88–92.

Watson, D. H. (1985). Toxic fungal metabolites in food. *CRC Crit. Rev. Fd Sci. Nutr.,* **22**, 177–98.

Wildman, J. D., Stoloff, L. & Jacobs, R. (1967). Aflatoxin production by a potent *Aspergillus flavus* Link isolate. *Biotechnol. Bioeng.,* **9**, 429–37.

Williams, B. C. (1985). Mycotoxins in foods and foodstuffs. In *Mycotoxins: a Canadian Perspective,* ed. P. M. Scott, H. L. Trenholm & M. D. Sutton. National Research Council Canada, Ottawa, pp. 49–53.

Wilson, D. M., Tabor, W. H. & Trucksess, M. W. (1976). Screening method for the detection of aflatoxin, ochratoxin, zearalenone, penicillic acid and citrinin. *J. Ass. Offic. Anal. Chem.,* **59**, 125–7.

Wilson, T., Rabie, C. J., Fincham, J. E., Steyn, P. S. & Schipper, M. A. A. (1984). Toxicity of rhizonin A, isolated from *Rhizopus microsporus,* in laboratory animals. *Fd Chem. Toxicol.,* **22**, 275–81.

Yazaki, H., Takahashi, H. & Nanayama, Y. (1980). Chemical detection and production of mycotoxins by *Penicillium verrucosum* var. *verrucosum* on stored meats. *Proc. Jap. Assoc. Mycotoxicol.,* no. 10, pp. 29–31 (in Japanese).

Yndestad, M. & Underdal, B. (1975). Aflatoxin in foods on the Norwegian market. *Nord. Vet.-Med.,* **27**, 42–8 (in Norwegian).

Yousef, A. E. & Marth, E. H. (1987). Quantitation of growth of mold on cheese. *J. Fd Protect.,* **50**, 337–41.

Zappavigna, R. & Cerutti, G. (1973). Le aflatossine nel latte e derivati: attuale conoscenze e primi risultati sperimentali. *Latte,* **47** 399–404.

Zerfiridis, G. K. (1985). Potential aflatoxin hazards to human health from direct mold growth on Teleme cheese. *J. Dairy Sci.,* **68**, 2184–8.

Zimmerli, B. (1977). Beitrag zur Bestimmung von Aflatoxinen mittels Hochdruck-Flüssigkeitschromatographie. *Mitt. Gebiete Lebens. Hyg.,* **68**, 36–45.

INDEX

Acremonium spp., toxins produced
by, 214
Activated charcoal, AFM_1
degradation affected by, 146
Adsorbosil® TLC plates, 69
Adsorption, AFM_1 affected by, 146
Aflatoxin B_1
biotransformation of, 6, 12, 24
carry-over into milk, 6, 12-16
chemical conversion of, 24
chemical structure of, 3, 4, 150,
195
degradation of, 151-2
diffusion in cheese, 224-5
legal limits in animal feedstuffs,
15, 32-40
frequency distribution of, 38
occurrence of, 3, 217-20, 238, 239
production in cheese, 230-1, 233
stability of, 227
Aflatoxin B_{2a}, 237
Aflatoxin G_1
diffusion in cheese, 224-5
formation in cheese, 220, 230-1,
233
stability of, 227
Aflatoxin M_1
absence in mould cultures, 166-8
analysis for
chromatographic methods, 57-90

Aflatoxin M_1—*contd.*
analysis for—*contd.*
immunochemical methods,
97-122
variation due to inadequate
methods, 142-3
antibody specificity for, 107-9
biosynthesis of, 24
carcinogenicity of, 26-31
risk assessment for, 29-30
chemical structure of, 6, 13, 58,
150
chemical treatment effects on,
144-5
chemistry of degradation of,
149-56
chromatographic methods of
analysis for, 57-90
clean-up procedures, 63-7
early developments, 58-9
inter-laboratory comparison,
86-7
sample extraction methods,
59-63
separation procedures, 67-85
validation procedures, 85-6
degradation of, 144-56
chemical treatment effects on,
144-5
chemistry of, 149-56

Aflatoxin M_1—*contd.*
  degradation of—*contd.*
    physical treatment effects on,
      146–9
    detection limits for
      chromatographic methods, 71,
        75, 80, 83
      immunochemical methods, 111,
        115, 116, 120, 122
      survey data affected by, 23
    first separated, 58
    fluorescence of, 12, 58, 70, 80
    identity confirmation of, 75–8
    immunochemical methods of
      analysis for, 97–122
      antibody production methods,
        101–7
      comparison of procedures,
        120–2
      determination procedures,
        109–20
      protocols used, 100
      types of, 98–100
    legal limits in dairy products, 39,
      40–1, 42–5
      frequency distribution of, 39
    occurrence in milk products,
      16–23
      surveys (1970s), 17, 18
      surveys (1980s), 17, 19–23
    physical treatment effects on,
      146–9
    reference materials, 41, 46, 47, 48,
      69, 88
    relationship to aflatoxin B_1 levels,
      13–14, 15
    stability during processing,
      128–42
      sources of variability, 142–3
    storage behaviour of, 69,
      132–3
    synthesis of, 24–5
    toxic effects of, 23–31
    toxicity studies, 25–6
      synthesis of material for,
        24–5
    UV radiation effects on, 146–9,
      154–6

Aflatoxin M_2
  chemical structure of, 6, 13, 58,
    150
  first separated, 58
Aflatoxin M_{2a}
  chemical structure of, 150
  formation of, 151, 152
Aflatoxin M_4, 12
  chemical structure of, 13
Aflatoxin M_x, formation of, 154–6
Aflatoxins
  analytical determination of,
    57–122, 240, 241, 242–4
  carry-over into dairy products
    high aflatoxin levels, 12–14
    low aflatoxin levels, 14–16
  detection limits for, 71, 75, 80, 83,
    111, 115, 116, 120, 122, 243
  first discovered, 2
  migration in cheese, 224–5
  production in cheese, 217–20
  production in dairy products,
    236–7
  reference materials for inter-
    laboratory comparison, 41, 46
  regulations for, 3, 31–47
  separation of, 2
  stability of, 227
Alimentary Toxic Aleukia (ATA), 1–2
Alternaria spp., contamination by,
  198–200, 212, 234, 235
Ames mutagenicity test, 28, 168, 175
Ammonia treatment, 127
Analytical methods
  chromatographic methods, 57–90
  comparison of, 84–5, 88, 120–1
  immunochemical methods, 97–122
Animal feedstuff ingredients
  aflatoxins in, 2, 16, 127–8
  reference materials, 46, 47, 48
Animal feedstuffs
  aflatoxin legal limits, 15, 32–40
    frequency distribution of, 38
Antibody
  meaning of term, 104–5
  monoclonal, 105–7
  polyclonal, 104–5
  specificity of, 107–9

Antifungal agents, 220–1
Antigens, 98
Argentina, dairy products AFM₁ limits, 42
Aspergillus spp.
  A. candidus, 207, 212
  A. flavus
    AFB₁ synthesized by, 2
    AFM₁ synthesized by, 24, 69
    contamination by, 207, 210–11, 234, 235, 236, 237
    occurrence in cheese, 207, 217–20
    temperature limits for growth of, 219–20
  A. fumigatus, 207, 212
  A. glaucus, 207, 212, 235
  A. nidulans, 217
  A. niger, 207, 212
  A. ochraceus, 203
  A. oryzae, 220
  A. parasiticus
    AFB₁ synthesized by, 2
    AFM₁ synthesized by, 24
    AFM₄ synthesized by, 12
    contamination by, 207, 217–20, 236, 237
    temperature limits for growth of, 219–20
  A. ustus, 207, 212
  A. versicolor, 7, 206–7, 208–9
    toxins produced by, 206, 207, 208–9, 217
    cheese contaminated by, 198–200, 206–12
    contamination by, 206–12, 234, 235
    dairy products contamination by, 236–7
    other toxigenic species, 207, 212
    toxins produced on cheese, 206, 207, 208–9, 217–20, 230–1, 233
Association of Official Analytical Chemists (AOAC) methods, 41, 59, 61, 64, 66, 77, 78
Austria
  animal feedstuffs aflatoxin limits, 33
  dairy products AFM₁ limits, 42
  milk survey, 19

Balkan endemic nephropathy, 3
Belgium
  dairy products AFM₁ limits, 43
  milk surveys, 18, 19
Benzoyl peroxide, AFM₁ affected by, 149
Bisulphite treatment, 128, 144
Bisulphites, AFM₁ affected by, 128, 144, 152
Blue-veined cheeses, 163, 165
  moulds used, 165
  see also Roquefort
Bovine serum albumin (BSA), 103
Brazil
  animal feedstuffs aflatoxin limits, 33
  dairy products AFM₁ limits, 43
Brick cheese
  AFM₁ in, 137, 139
  Aspergilli growth on, 219
  diffusion of aflatoxins in, 225
Brie cheese, 164, 178
  toxins in, 182
Butter
  fungal contamination of, 235
  manufacture of, 134–5
Buttermilk, fungal contamination of, 236

Camag Linomat® spotting device, 75
Camembert cheese, 164, 178
  AFM₁ in, 137, 139
  toxins in, 182
Canada, animal feedstuffs aflatoxin limits, 33
Carcinogenicity studies, 26–31
Carry-over (of mycotoxins), 6–7, 11–16
  cheese-making, 137, 138–9
  high aflatoxin levels, 12–14
  low aflatoxin levels, 14–16
Casein
  aflatoxin production affected by, 218
  association of AFM₁ with, 137, 140–1
  proteolysis of, 142

Celite® extraction columns, 61
Cellulose columns, 64
Cephalosporium spp., toxins
    produced by, 214
Cheddar cheese, 131
    AFM_1 in, 137, 139
    Aspergilli growth on, 219
    diffusion of aflatoxins in, 224
    fungal contamination of, 198, 209,
        210, 215
    packaging of, 215
    toxins produced in, 180, 198, 204
Cheese
    AFM_1 in, 23, 136–7, 138–9
    carry-over of AFM_1 into, 137,
        138–9
    health risk of metabolites in,
        185–7
    heat treatment of, 141
    manufacture of, 135–7
        moulds used, 164–5
        stability of AFM_1 during, 136–7
    melting of, 227
    migration of mycotoxins in, 222–5
    mycotoxins produced on, 178–83,
        214–21
    packaging of, mould growth
        affected by, 214–15
    penetration of mycotoxins in,
        222–6
    prevention of toxin formation in,
        184–7
    ripening conditions to minimize
        toxin production, 184–5
    ripening of, 142
    storage conditions to minimize
        toxin production, 185
    toxic metabolites in, 178–83,
        214–21
        commercial cheeses, 181–3
        detection methods used, 178,
            179
        experimental cheeses, 178, 180–1
    trimming of, 226
    washing of, 227
    see also Blue-veined . . . ;
        Mould-ripened . . . ; White
        surface mould cheese

Cheesecake, fungal contamination
    of, 211
ChemElut® extraction columns, 61
Chemical treatments, AFM_1 affected
    by, 144–5
China, milk survey, 20
Chromatographic methods
    clean-up procedures for, 63–7, 241,
        243
    detection limits of, 71, 75, 80, 83
    early developments, 58–9
    inter-laboratory comparison of,
        86–7
    sample extraction for, 59–63
    separation procedures for, 67–85
    validation procedures for, 85–6
    see also High-performance
        liquid . . . ; Thin-layer
        chromatography
Citreoviridin, 217
Citrinin
    analytical determination of, 240–1
    chemical structure of, 3, 4, 195
    migration in cheese, 222–3
    production in cheeses, 201, 202–3,
        215–16, 229
Cladosporium spp.
    C. butyri, 235
    contamination by, 198–200, 212,
        234, 235
Clean-up procedures, 63–7
    column chromatography, 64–6
    immunoaffinity cartridges used,
        66
    silica gel cartridges used, 65–6
    solvent partition techniques, 63–4
    thin-layer chromatography, 63
Colby cheese, AFM_1 in, 139
Cold storage, AFM_1 affected by,
    132–3
Collaborative studies,
        chromatographic methods,
        85–6
Column chromatography
    purification using, 64–6
    sample extraction using, 62–3
    separation by, 68
Condensed milk, aflatoxins in, 237

Consumer complaints, mould in
food, 194
Cottage cheese
aflatoxins formed in, 220
AFM$_1$ in, 136, 138
Aspergilli growth on, 219, 220
fungal contamination of, 211
Coumarin, chemical structure of,
150
Covalent binding index (CBI), 27–8
Cream
aflatoxins in, 236
fungal contamination of, 235
manufacture of, 134–5
Cream cheese, fungal contamination
of, 206, 217
Cultured dairy products,
manufacture of, 133
Cyclopiazonic acid, 168, 172–3
analytical determination of, 169,
179
chemical structure of, 172
production in cheese, 182, 184,
185, 187, 207, 229
production in milk, 237
toxicity of, 172, 186
Cylindrocarpon spp., toxins produced
by, 214
Czechoslavakia, dairy products
AFM$_1$ limits, 43

Dairy products
aflatoxin regulations for, 39, 40–1,
42–5
current limits, 41, 42–5
factors affecting criteria, 40–1
AFM$_1$ affected by processing,
128–43
component products, 133–42
factors affecting variability in
results, 142–3
whole-milk processing, 129–33
contamination by mycotoxins,
5–7
direct contamination of, 7
fungal contamination of, 234–9
indirect contamination of, 6–7

Damietta cheese, fungal
contamination of, 209, 211
Deoxynivalenol
chemical structure of, 3, 4
effect of, 3, 5
Derivatization
chromatographic methods, 82
immunochemical methods, 101–3
Detection limits
cheese toxins, 173, 174, 175, 176,
177, 179, 240, 242, 243
chromatographic methods, 71, 75,
80, 83, 173, 174, 175, 176, 177,
179, 240, 242, 243
immunochemical methods, 111,
115, 116, 120, 122
survey data affected by, 23
Diatomaceous earth columns, 61, 64
Direct contamination, 7
Dominican Republic, animal
feedstuffs aflatoxin limits, 33
Drechslera spp., cheese contaminated
by, 213–14
Dried milk products, fungal
contamination of, 234–5
Ducklings, toxicity studies on, 6, 25,
58

Easi-extract® columns, 119
Edam cheese, fungal contamination
of, 199, 203, 206, 208, 210, 215,
217, 230
Emmental cheese, 131
Enzyme immunoassay (EIA), 111–16
AFM$_1$ in milk determined by,
112–16
detection limits of, 115–116
principle of, 111–12
Enzyme-linked immunosorbent
assay (ELISA), 40, 111–16, 117
advantages of, 121
detection limits of, 115
direct method, 112–13, 115
indirect method, 112, 114, 116–117
Epidemiological studies, 30
Ergot, regulations on, 3
Ergotamine, chemical structure of, 4

Ergotism, 1
European Community
  animal feedstuffs aflatoxin limits,
    34
  Bureau of Reference (BCR)
    intercomparison studies, 86, 87
  Mycotoxin Programme, 46
Evaporated milk, aflatoxins in, 239
Evaporation, AFM$_1$ affected by, 130,
    134
Extraction methods, 59–63
  solid-phase extraction methods,
    62–3
  solvent extraction methods, 60–2
Extrelut® extraction columns, 62

Faecilomyces spp., contamination by,
    235
Feedstuffs. *See* Animal feedstuffs
Fermented milk products
  fungal contamination of, 235–6
  see also Buttermilk; Yoghurt
Feta cheese
  aflatoxins removed from, 227
  Aspergilli growth on, 219
  diffusion of aflatoxins in, 225
Finland, milk survey, 20
Florisil® column, 242, 243
Fluorescence, AFM$_1$, 12, 58, 70, 80
Fluorescent materials, AFM$_1$
    analysis affected by, 143
Food and Agriculture Organization
    (FAO), legislation survey, 31
France
  animal feedstuffs aflatoxin limits,
    34
  dairy products AFM$_1$ limits, 43
  milk survey, 20
Freeze-drying, AFM$_1$ affected by,
    134
Frozen storage, AFM$_1$ affected by,
    132–3
Fungal cheese starter cultures
  low-toxicity strains used, 184
  toxic metabolites from, 165–83
    biosynthetic pathways for,
    166

Fungal cheese starter cultures—*contd.*
  toxic metabolites from—*contd.*
    detection methods used, 168,
    178
    identification in cheeses, 178–
    83
    identification *in vitro*, 168–78
Fungal contaminants, 193–214
  toxins produced by, 214–21
    occurrence in mouldy cheese,
    227–34
    penetration in cheese, 222–5
    stability in cheese, 226–7
Fungal growth, prevention of, 220–1
Fungal processed cheeses, 163–4
  see also Mould-ripened cheese
Fusarium spp., contamination by,
    198–200, 212, 234, 235

Genotoxicity tests, 27–8
Geotrichum spp., contamination by,
    212–13, 234, 235
Germany (Democratic Republic),
    milk surveys, 17, 18
Germany (Federal Republic)
  dairy products AFM$_1$ limits,
    43
  milk surveys, 18, 21
Gliocladium spp., toxins produced
    by, 214
Gliotoxin, 214
Gouda cheese
  AFM$_1$ in, 137, 139
  fungal contamination of, 199, 206,
    208, 210, 215, 230
Grated cheese, mould growth on,
    220
Gruyère cheese, fungal
    contamination of, 210

Haptens, 98
Health risk, cheese toxins,
    185–7
Heat treatments
  cheese, 141
  milk, 129–32

High-performance liquid
  chromatography (HPLC), 40,
  41, 78–85, 243
  detection limits of, 80, 83
  identity confirmation in, 83–4
  immunoaffinity chromatography
    combined with, 119–20, 122
  normal-phase, 79, 80–1, 243
    detection limits of, 80
  principle of, 78–80
  reverse phase, 79, 81–3
    detection limits of, 83
Hydrogen peroxide, AFM_1 affected
  by, 144–5, 148–9, 152–4

Immunization procedures, 104
Immunoaffinity cartridges
  clean-up using, 66
  see also Enzyme-linked
    immunosorbent assay
    (ELISA)
Immunoaffinity chromatography
  (IAC), 116–20
  combined with HPLC, 119–20
  combined with
    radioimmunoassay, 119
  detection limit of, 120, 122
  purification using, 99
Immunochemical methods
  antibody production procedures,
    102–7
  comparison of procedures, 120–2
  detection limits of, 111, 115, 116,
    120, 122
  enzyme immunoassay, 111–16
  immunoaffinity chromatography,
    116–20
  procedures, 109–20
  protocols for, 100
  radioimmunoassay, 109–11
  types of, 98–100
Immunocomplex precipitation
  methods, 99
Immunogen
  preparation of, 101–4
    conjugation to protein, 103–4
    derivatization used, 101–3

Immunoglobulins, 104–5
India
  animal feedstuffs aflatoxin limits,
    34
  milk surveys, 17, 18
Infant foods
  aflatoxins in, 237
  AFM_1 limits for, 42, 43, 44, 45
Inter-laboratory comparison
  chromatographic methods, 86–7
  reference materials for, 41, 46
International Agency for Research
  on Cancer (IARC), Check
  Sample Programme, 87
International Dairy Federation
  (IDF) methods, 59, 61, 64
International Union of Pure and
  Applied Chemistry (IUPAC)
  methods, 59, 61, 64, 77
Ireland, milk survey, 21
Isofumigaclavines, 177, 182, 183
  analytical determination of, 169,
    177, 179
  chemical structure of, 177
  pharmacological action of, 177
  production in cheese, 182, 183,
    187
Israel, animal feedstuffs aflatoxin
  limits, 34
Italy, milk survey, 21
Ivory Coast, animal feedstuffs
  aflatoxin limits, 35

Japan, animal feedstuffs aflatoxin
  limits, 35
Jordan, animal feedstuffs aflatoxin
  limits, 35

Kachkaval cheese, fungal
  contamination of, 211
Kefir, manufacture of, 133
Khoa, aflatoxins in, 134, 237

Lactoperoxidase, AFM_1 degradation
  affected by, 145, 153

Legislation
  aflatoxins, 3, 31–47
  animal feedstuffs, 31–40
  dairy products, 40–1, 42–5
  reference materials used, 41, 46,
    47, 48
Low-temperature storage
  AFM$_1$ affected by, 132–3
  toxic metabolites formed during,
    185
Luteoskyrin, 201, 206

Milk
  carry-over of aflatoxins into, 6–7,
    11–16
    reaction time for, 13
  chemical treatment of, 144–5
  concentration of, 133–4
  freeze-drying of, 134
  fungal contamination of, 234
  heat treatment of, 129–32
  low-temperature storage of, 132–3
  physical treatment of, 146–8
Milk powder
  aflatoxins in, 236–7, 238, 239
  fungal contamination of, 234–5
  manufacture of, 130, 133–4
  ochratoxin A in, 238
  reference materials, 46, 47, 48
Milk products
  aflatoxins in, 236–7, 239
  AFM$_1$ in, 16–23
    surveys (1970s), 17, 18
    surveys (1980s), 17, 19–23
  fungal contamination of, 234–6
Monoclonal antibody
  production of, 105–7
  specificity of, 107–9
Mould cultures. *See* Fungal cheese
    starter cultures
Mould-ripened cheese
  Aspergilli growth on, 218–19
  moulds used, 164–5
  potentially toxic metabolites from,
    165–83
  see also Blue-veined . . .; White
    surface mould cheese

Mouldy cheese, mycotoxins in,
    227–34
Mozzarella cheese
  AFM$_1$ in, 137, 139
  heat treatment of, 141
Mucor spp., contamination by, 213,
    234, 235
Multimycotoxin analytical methods,
    240–1
Mutagenicity tests, 28, 168, 175
Mycophenolic acid, 176–7
  analytical determination of, 169,
    176, 179, 240
  antibiotic properties of, 176
  chemical structure of, 176
  health risks of, 186
  production in cheeses, 181, 182,
    183, 185, 201, 228–9
Mycotoxins, significance of, 1–5

Natamycin, 221
Netherlands
  dairy products AFM$_1$ limits, 44
  milk surveys, 18, 22
Nigeria
  animal feedstuffs aflatoxin limits,
    35
  dairy products AFM$_1$ limits, 43
β-Nitropropionic acid
  analytical determination of, 244
  production in cheese, 220, 231, 233
Norway, animal feedstuffs aflatoxin
    limits, 35

Ochratoxin A
  analytical determination of, 240–1
  chemical structure of, 3, 4, 195
  migration in cheese, 222–3
  occurrence of, 3, 215–16
  production in cheeses, 201, 202–3,
    215–16, 229
  stability of, 227
Octyldecylsilane-bonded silica gel
  clean-up using, 65–6, 82
  sample extraction using, 62–3
  separation using, 81–3

Oman, animal feedstuffs aflatoxin
limits, 35
Oospora spp., toxins produced by,
212, 214
Oosporein, 214
Oxidizing agents, effects of UV
radiation affected by, 148–9

Parmesan cheese, AFM₁ in, 137, 139
Particulate materials, AFM₁
degradation affected by, 146
Pasteurization, AFM₁ affected by,
130, 131
Patulin, 173
analytical determination of, 169,
173, 179, 240
chemical structure of, 3, 4, 172,
195
instability of, 180
migration in cheese, 223
occurrence of, 3, 216
production in cheeses, 180, 185,
201, 204–5, 216, 228
stability of, 226–7
toxicity of, 173
Penicillic acid, 173–4
analytical determination of, 169,
174, 240, 241
chemical structure of, 174, 195
migration in cheese, 223
production in cheeses, 180, 185,
201, 204–5, 216, 228
stability of, 226–7
toxicity of, 174
Penicillin, production of, 194
Penicillium spp., 196–206
classification of, 197
contamination by, 196–206, 234,
235
P. aurantiogriseum, 203, 215
P. brevicompactum, 197, 201
P. *camemberti*, 164–5
synonyms for, 164
toxins produced by, 168, 172–3
P. citreoviride, 217
P. citrinum, 216
P. claviforme, 216

Penicillium spp.—*contd.*
P. *commune*, 202, 215
P. *crustosum*, 197, 206, 217, 232
P. *cyclopium*, 196–7, 201, 203, 206,
215, 216
P. *expansum*, 205
P. *patulum*, 216, 223
P. *puberulum*, 216, 236
P. *roqueforti*, 165
synonyms for, 164
toxins produced by, 173–8, 204,
216
P. *stoloniferum*, 176
P. *urticae*, 205, 216
P. *verrucosum* var. *cyclopium*,
196–7, 201, 203, 206, 237
P. *verrucosum* var. *verrucosum*, 215
P. *viridicatum*, 197, 201, 203, 215,
216
toxins produced by, 7, 168, 172–8,
201, 215–17, 228–9, 232–3
Penitrem A, production on cheese,
206, 217, 229
Peru, animal feedstuffs aflatoxin
limits, 35
pH, effects of UV radiation affected
by, 148
Phoma spp., contamination by, 235
Physical treatments, AFM₁ affected
by, 146–9
Pimarcin, 221
Poland, animal feedstuffs aflatoxin
limits, 35
Polyclonal antibody
production of, 104
purification of, 104–5
specificity of, 107, 108
Portugal, animal feedstuffs aflatoxin
limits, 35
PR-imines, toxicity of, 175
PR-toxin, 174–6
chemical structure of, 175
toxicity of, 175
Processed cheese, *Aspergilli* growth
on, 220
Provolone cheese
diffusion of aflatoxins in, 224
fungal contamination of, 209

Queso Blanco cheese
AFM₁ in, 138
milk processed for, 130

Radioimmunoassay (RIA), 109–11
aflatoxins in dairy products
determined by, 110–11, 244
detection limit of, 111, 122
disadvantages of, 121
immunoaffinity chromatography
combined with, 119
incubation step in, 109
principle of, 109–10
radioactivity measurements for,
109–10
separation step in, 109
Rainbow trout, carcinogenicity
studies using, 26–7
Rats
carcinogenicity studies using, 28–9
toxicity studies on, 25–6
Reference materials, inter-laboratory
comparison, 41, 46, 88
Refrigerated stores, AFM₁ affected
by, 132–3
Refrigerators, air temperature in, 220
Regulations, 31–47
aflatoxins, 3, 31–47
animal feedstuffs, 31–40
current limits, 32–40
factors affecting criteria, 31–2
dairy products, 40–1, 42–5
current limits, 41, 42–5
factors affecting criteria, 40–1
ergot, 3
reference materials used, 41, 46,
47, 48
Regulosin, 206
Retardation factor, definition of, 69
Reverse-phase liquid
chromatography
sample extraction using, 62
separation by, 81–3
Rhizoctonia spp., toxins produced by,
214
Rhizopus spp., contamination by,
213, 234, 235

Riboflavin
AFM₁ degradation affected by,
145, 153
photosensitization of, 153
Ricotta cheese
AFM₁ in, 138
heat treatment of, 141
milk processed for, 130
Roller-drying, AFM₁ affected by,
130, 134
Romania
animal feedstuffs aflatoxin limits,
36
dairy products AFM₁ limits, 44
Roquefortine, 177–8, 182, 183
analytical determination of, 169,
177, 179
chemical structure of, 177
production in cheese, 182, 183,
184, 185, 187
toxicity of, 177
Russia. *See* USSR

Saint-Nectaire cheese, fungal
contamination of, 198, 211,
212
Sampling problems, 32, 59–60
Scopulariopsis spp., contamination
by, 235
Senegal, animal feedstuffs aflatoxin
limits, 36
Separation procedures,
chromatography, 67–85
Sepharose® columns, 66, 117,
119
Sezet (antifungal agent), 221
Silica gel cartridges
clean-up using, 65–6
sample extraction using, 62–3
Silica gel columns
purification using, 65–6
sample extraction using, 62–3
separation using, 79, 80–1
Slaframe, 214
Solid-phase extraction columns,
samples extraction using,
62–3

Solvent partition techniques
  clean-up using, 63–4
  sample extraction using, 60–2
South Africa, milk survey, 18
Spain, milk survey, 22
Spray-drying, AFM$_1$ affected by, 130, 134
Steam injection, AFM$_1$ affected by, 131
Sterigmatocystin
  analytical determination of, 240, 241–2
  chemical structure of, 3, 4, 195
  migration in cheese, 223–4
  production in cheese, 207, 208–9, 217, 230
  stability of, 227
Sterilization, AFM$_1$ affected by, 130
Streptococcus lactis, 237
Sulphites, AFM$_1$ affected by, 144
Swainsonine, 214
Sweden
  animal feedstuffs aflatoxin limits, 36
  dairy products AFM$_1$ limits, 44
  milk survey, 22
Swiss cheeses
  AFM$_1$ in, 139
  Aspergilli growth on, 219
  fungal contamination of, 198, 209, 210
  toxins produced in, 180, 198, 202, 204
Switzerland
  animal feedstuffs aflatoxin limits, 36
  dairy products AFM$_1$ limits, 45

T-2 toxin, 3, 5
  chemical structure of, 4
Teleme cheese, 210
Temperature
  Aspergilli growth affected by, 219–20
  effects of UV radiation affected by, 147–8

Thin-layer chromatography (TLC), 40, 41, 68–78
  cheese toxic metabolites detected by, 168, 178, 179, 240–4
  cyclopiazonic acid, 169, 179
  detection limits of, 71, 75, 173, 174, 175, 176, 177, 179, 240, 241, 242, 243
  identity confirmation for, 75–8
  isofumigaclavines, 169, 177, 179
  mycophenolic acid, 169, 176, 179, 240
  one-dimensional, 70–1
    detection limit for, 71
    spotting pattern for, 70
  other procedures, 74–5
  patulin, 169, 173, 179, 240, 241
  penicillic acid, 169, 174, 179, 240, 241
  PR toxin, 169, 175, 179
  principle of, 68–9
  roquefortine, 169, 177, 179
  spotting patterns used, 70, 71–3, 74–5
  two-dimensional, 71–4
    detection limits for, 75, 240, 242
    plate size used, 73
    spotting patterns for, 71–3, 74, 75
Tilsit cheese
  AFM$_1$ in, 137, 139
  Aspergilli growth on, 219, 220
  diffusion of aflatoxins in, 224, 225
  fungal contamination of, 216, 217, 231
  washing of, 227
Tome de Savoie cheese, fungal contamination of, 198, 212
Toxicity studies, AFM$_1$, 25–6
Trichoderma spp., toxins produced by, 214
Trichothecenes, 3, 214
Trifluoroacetic acid (TFA)
  AFM$_1$ identified using, 76–8
  derivatization in HPLC, 82
  sterigmatocystin identified by, 242
Turkey X Disease, 2

UK
  animal feedstuffs aflatoxin limits,
    36
  milk surveys, 18, 22
Ultraviolet (UV) radiation
  AFM_1 affected by, 146–9, 154–6
    effect of oxidizing agents, 148–9
    pH effects, 148
    temperature effects, 147–8
Uruguay, animal feedstuffs aflatoxin
    limits, 37
USA
  animal feedstuffs aflatoxin limits,
    37
  dairy products AFM_1 limits, 45
  milk survey, 18
USSR, dairy products AFM_1 limits,
    45

Versicolorin A, 207
Verticillin A, 214
Verticillium spp.
  contamination by, 235
  toxins produced by, 214

Vomitoxin. *See* Deoxynivalenol

Whey, AFM_1 in, 137, 140
White surface mould cheese, 164
  moulds used, 164–5
  see also Brie; Camembert
Wilstermarsch cheese
  fungal contamination of, 217
  migration of toxins in, 223

Yeast extract–sucrose (YES), 166, 167
Yellow rice, 2
Yog(h)urt
  aflatoxins in, 237
  fungal contamination of, 235–6
  manufacture of, 133
Yugoslavia, animal feedstuffs
    aflatoxin limits, 37

Zearalenone, 5
  chemical structure of, 4